AF767198

SMART CITY 2.0

Strategies and Innovations for City Development

World Scientific Series in R&D Management

Print ISSN: 2591-7498
Online ISSN: 2591-7501

Series Editor: Tugrul U Daim *(Portland State University, USA)*

Published

Vol. 8 *Smart City 2.0: Strategies and Innovations for City Development*
edited by Deog-Seong Oh, Fred Young Phillips and Avvari V Mohan

Vol. 7 *A Perspective on Holistic Engineering Management: Learning, Adapting and Creating Value*
by Robert J Aslett, John M Acken and Siva K Yerramilli

Vol. 6 *Digital Transformation: Evaluating Emerging Technologies*
edited by Tugrul U Daim

Vol. 5 *Managing Medical Technological Innovations: Exploring Multiple Perspectives*
edited by Tugrul U Daim and Alexander Brem

Vol. 4 *Managing Mobile Technologies: An Analysis from Multiple Perspectives*
edited by Tugrul U Daim and Alexander Brem

Vol. 3 *Cooperative Innovation: Science and Technology Policy*
by Frederick Betz

Vol. 2 *Technology Roadmapping*
edited by Tugrul U Daim, Terry Oliver and Rob Phaal

Vol. 1 *Managing Technological Innovation: Tools and Methods*
edited by Tugrul U Daim

World Scientific Series in R&D Management – Vol. 8

SMART CITY 2.0

Strategies and Innovations for City Development

Editors

Deog-Seong Oh
Woosong University, South Korea

Fred Young Phillips
University of New Mexico, USA & Tongji University, China

Avvari V Mohan
University of Nottingham Malaysia Campus, Malaysia

World Scientific

EW JERSEY • LONDON • SINGAPORE • BEIJING • SHANGHAI • HONG KONG • TAIPEI • CHENNAI • TOKYO

Published by

World Scientific Publishing Co. Pte. Ltd.
5 Toh Tuck Link, Singapore 596224
USA office: 27 Warren Street, Suite 401-402, Hackensack, NJ 07601
UK office: 57 Shelton Street, Covent Garden, London WC2H 9HE

Library of Congress Cataloging-in-Publication Data
Names: Oh, Deog-Seong, editor. | Phillips, Fred, 1952– editor. | Avvari V. Mohan, editor.
Title: Smart city 2.0 : strategies and innovations for city development / editors Deog-Seong Oh, Woosong University, South Korea, Fred Young Phillips, University of New Mexico, USA & Tongji University, China, Avvari V Mohan, University of Nottingham Malaysia Campus, Malaysia.
Description: Singapore ; Hackensack, NJ, USA : World Scientific Publishing Co. Pte. Ltd., [2023] | Series: World scientific series in R&D management, 2591-7498 ; vol. 8 | Includes bibliographical references and index.
Identifiers: LCCN 2022025784 | ISBN 9789811257179 (hardcover) | ISBN 9789811257186 (ebook for institutions) | ISBN 9789811257193 (ebook for individuals)
Subjects: LCSH: Smart cities.
Classification: LCC TD159.4 .S637 2023 | DDC 307.760285--dc23/eng/20220801
LC record available at https://lccn.loc.gov/2022025784

British Library Cataloguing-in-Publication Data
A catalogue record for this book is available from the British Library.

Copyright © 2023 by World Scientific Publishing Co. Pte. Ltd.

All rights reserved. This book, or parts thereof, may not be reproduced in any form or by any means, electronic or mechanical, including photocopying, recording or any information storage and retrieval system now known or to be invented, without written permission from the publisher.

For photocopying of material in this volume, please pay a copying fee through the Copyright Clearance Center, Inc., 222 Rosewood Drive, Danvers, MA 01923, USA. In this case permission to photocopy is not required from the publisher.

For any available supplementary material, please visit
https://www.worldscientific.com/worldscibooks/10.1142/12871#t=suppl

Desk Editors: Jayanthi Muthuswamy/Amanda Yun

Typeset by Stallion Press
Email: enquiries@stallionpress.com

© 2023 World Scientific Publishing Company
https://doi.org/10.1142/9789811257186_fmatter

Foreword

Much has been written and is being written about smart cities. This suggests, first, that there are no simple answers and, second, that agreement about the topic is not yet total. The editors of this volume wished not to add to the cacophony but rather to help focus it. In this they've succeeded, by gathering chapters that re-examine the driving purposes of smart cities. As smart city projects proliferate worldwide, cultural differences and preferences, so difficult to capture in artificial intelligence, must be respected. The editors of this volume represent three countries, and the chapter authors offer international perspectives from four continents. These authors are authorities in technology-based economic development and city planning experts. They propose additional technologies for the smart city mix. They urge readers to view smart cities as balancing efficiency, quality of life and entrepreneurial potential.

Kingsley E. Haynes
Dean Emeritus, Schar School of Policy and Government,
George Mason University Fellow,
National Academy of Public Administration Fellow,
Western Regional Science Association President Emeritus,
Regional Science Association International.

© 2023 World Scientific Publishing Company
https://doi.org/10.1142/9789811257186_fmatter

About the Editors

Deog-Seong Oh is the President of Woosong University and a former President of Chungnam National University, both in Daejeon, South Korea. Dr. Oh is an expert in public and urban policy, economic development and technology commercialization.

Dr. Oh is a member of the UNESCO High Panel on Science for Development and Organizing Committee Chair for UNESCO's Global Innovation Forum. He is a former Secretary- General of World Technopolis Association. He received a Master of Urban Planning in 1979 and an M.S. in Architecture from Seoul National University, Korea, in 1981. He earned a Ph.D. in Urban Planning from Hanover University, Germany, in 1989. He did post- doctoral research at the University of Sheffield, UK, in 1993, and was visiting Professor at the University of Dortmund, 2002–2004.

He was the Vice President of the Korean Planners Association and Korean Urban Management Association. He is also the chief editor of *Asian Pacific Planning Review*. He has published 250 papers on urban planning and design, sustainable development and regional innovation.

Fred Young Phillips is currently on faculty at the State University of New York at Stony Brook in the USA and at Tongji University in Shanghai. He is the 2017 winner of the Kondratieff Medal, awarded by the Russian Academy of Sciences. He is the Vice-President of ACIEK, a society of innovation and entrepreneurship scholars, and he coordinates TANDO, a think tank recently spun out of the University of Texas at Austin. Fred is a Fellow of PICMET, the Portland International Center for Management of Engineering and Technology.

Dr. Phillips is the Editor-in-Chief Emeritus of Elsevier's international journal *Technological Forecasting & Social Change*. He has consulted worldwide on technology-based regional development. He is a founder of the Austin Technology Council and was also a Board member of the Software Association of Oregon. Dr. Phillips attended the University of Texas and Tokyo Institute of Technology, earning a Ph.D. at Texas (1978) in mathematics and management science.

Avvari V Mohan is the Professor and Dy Head of Engagement & Impact at Monash University Business School (Malaysia). Mohan received his doctorate in Marketing and Innovation from the Department of Management Studies of the Indian Institute of Science (IISc), Bangalore, following which he visited South Korea on a Research Fellowship at the Korea Advanced Institute of Science and Technology (KAIST). Prior to joining Monash University, he was the Associate Professor of Strategy & Innovation and the Director of Research at the Nottingham University Business School (NUBS) in the University of Nottingham Malaysia (UNM). He also served as a member of the Faculty of Management at Multimedia University, Cyberjaya, Malaysia.

His research interests are in the areas of Strategy and Innovation with special interest in sustainability-oriented Strategies. He is particularly interested in collaborations or linkages firms develop with other organizations for Innovation and Sustainable Development. He has served as a Council Member of the Consumer Forum (CFM) for the Malaysian Communications and Multimedia Industry, as a recourse person in WTA-UNESCO workshops and recently as Innovation Auditor for MIGHT, Malaysia.

© 2023 World Scientific Publishing Company
https://doi.org/10.1142/9789811257186_fmatter

Contents

© 2023 World Scientific Publishing Company
https://doi.org/10.1142/9789811257186_fmatter

List of Figures

© 2023 World Scientific Publishing Company
https://doi.org/10.1142/9789811257186_fmatter

List of Tables

© 2023 World Scientific Publishing Company
https://doi.org/10.1142/9789811257186_fmatter

List of Abbreviations

A

- AI - Artificial Intelligence
- APIs - Application Programming Interfaces

B

- BRASSCOM - Brazilian Association of Information and Communication Technology Companies

C

- CC - Competence Centre
- CEMS - Community Energy Management System
- CEREAL - Creativity, Exploration, Resilience, Experimentation, Adaptability and Learning to learn
- CNO - Collaborative Networked Organization

D

- DIGI+ - Development, Innovation, Governance, Inclusion, Upgrade
- DIH - Digital Innovation Hubs

E

- EDC - Eco Delta City
- ED - Economic Discrimination
- EDP - Entrepreneurial Discovery Process

G

- GARLICs - Government, Academia, Research Labs, Large enterprises, Investors, Citizens, and small and medium enterprises

H

- HCI - Heavy and Chemical Industry

I

- ICF - Intelligent Community Forum
- ICT - Information and Communication Technology
- IIRIS - Information Infrastructure for Regional Innovation Systems
- IoT - Internet of Things
- IPTBLS - Innopreneurship platform's triple bottom line score
- IS - Information System
- ITU - International Telecommunication Union

K

- KAIST - Korea Advanced Institute of Science and Technology
- KPIs - Key Performance Indicators

L

- LED - Light Emitting Diode
- LID - Low-Impact Techniques

M

- MaaS- Mobility as a Service
- MEC - Multi-access Edge Computing
- MMG - Minden Marketing GmbH
- MST - Ministry of Science and Technology (Korea)

N

- NGOs - Non-Governmental Organizations
- NIPS - Normalized Innopreneurship Platform's Score
- NTBF - New Technology-Based Firms

O

- OECD - Organisation for Economic Co-operation and Development

P

- PAC - Growth Acceleration Program (Programa de Aceleração do Crescimento)
- PDCA - Plan-Do-Check-Act cycle
- PISA - Programme for International Student Assessment
- PPP - Public–Private Partnerships
- PV - Photovoltaic

Q

- QOL - Quality Of Life

R

- RBV - Resource-Based View
- RIS - Regional Innovation System
- RES - Reengineered Economy Score

S

- SC - Smart City
- SCIPS - Smart City Innovation Platform's score
- SCSE - Smart City Summit & Expo
- SSC - Smart Sustainable City
- SMEs - Small- and Medium-Sized Enterprises
- STPs - Science and Technology Parks

T

- TBL - Triple Bottom Line
- TCA - Taipei Computer Association
- TUM - Technical University of Munich

U

- UAV - Unmanned Aerial Vehicle
- UFG - Federal University of Goiás
- UIG - University–Industry–Government

V

- VUCA - Volatility, Uncertainty, Complexity, and Ambiguity

© 2023 World Scientific Publishing Company
https://doi.org/10.1142/9789811257186_fmatter

Introduction

It is now 95 years—hard to believe!—since Fritz Lang's pioneering film *Metropolis*.[1] Lang's city is technology-heavy, both in its visible futuristic cityscapes and in its premise of hidden machines that keep the city humming and keep their proletariat attendants busy. These workers are the city's "hands". The city's overlords/designers are its "head". The film's female lead agitates for a mediating "heart" that will lead the city to a more humane regime.

What to make of the fact that, despite the passage of a century and attending advances in technology, we have made so little progress toward creating a city with "heart"? And that many people still identify urban technology with a Lang-like dystopia?[2] Several of the present volume's authors address this conundrum.

Technopolis and Smart City

The book was inspired by presentations at the 2019 UNESCO Global Innovation Forum (GIF 2019), though it includes chapters by scholars and practitioners who did not attend the GIF. Because GIF has succeeded the now-defunct World Technopolis Association, many GIF presenters have backgrounds in science/technology parks (STPs), and innovation ecosystems.[3] Though there is considerable tension between the STP orientation and a smart city orientation, the former contributes

[1]URL: https://www.imdb.com/title/tt0017136/.

[2]F. Phillips, (2006). *Social Culture and High Tech Economic Development: The Technopolis Columns*, p. 200. Palgrave Macmillan.

[3]D.S. Oh and F. Phillips (eds.) (2014). *Technopolis: Best Practices for Science & Technology Cities*. Springer.

helpfully to the latter. This is because smart city cannot rest exclusively on information technologies and the Internet of Things (IoT)—despite much literature that seems to presume otherwise.

STP-oriented authors herein write of the modern world's resource constraints. Advances in science and technology (physical, chemical, biological, geological, etc.) ease these constraints by discovering new resources, and by finding ways to use known resources more efficiently. Information and communication technologies (ICTs) provide only the second of these functions. (They support the first only indirectly.) The sustainable smart city will benefit from every new non-ICT technology from pharmaceutical advances to asteroid mining.

The 2019 UNESCO Global Innovation Forum featured leaders of STPs and smart city initiatives on the same panel. This proved not as explosive as moderator Fred Phillips had feared, though the potential was evident.

In some ways, the two different movements started in the same way. Companies like Bechtel had seen the science city movement as an opportunity to profitably pour concrete. The resulting metros, lacking human amenities (heart), turned into ghost cities. (Tsukuba's early days provided the prime example.) Bechtel subsequently developed an appreciation for public art and other means of making the second-generation technopolis livable.

Likewise, Cisco and others perceived the smart city idea as a profitable chance to lay cable. The result? Computers knew how to reside in the smart city, but people hesitated to do so. What will make Smart City 2.0 attractive and livable?

Technopolis and smart cities are very different, and indeed have opposing objectives. Smart cities are about efficiency and sustainability; technopolis and STPs are about innovation. Innovation is messy, wasteful (in the short run) and inefficient, as innovators try successive wild but plausible and ultimately unworkable ideas. A still newer urban concept, the eco-city, is likewise incompatible with the technopolis, as a newly launched high-tech product may offer enormous consumer benefit well before its maker figures out how to fabricate it in a greener way and recycle it more completely. The product thus would be deemed unsuitable for the eco-city. The smartphone offers food for thought in this regard.

"Pure" technopolis, smart cities, or eco-cities cannot be realized. Each must leave room for human muddling.

Data, Governance, and Customer-Orientation

To carry this argument farther, an eco-city may not be a sustainable smart city, if a lack of heart leads to cultural decline. Though definitions of technopolis, smart city, and eco-city are still subject to argument, they do seem different from and in some ways antithetical to each other. Authors in this volume attempt to reconcile them.

These differences evince the same tension that corporations experience in simultaneously pursuing creative exploration and efficient exploitation. Yet that tension is more difficult to manage in cities than in companies, because only some of a city's multiple constituencies are city employees. The remaining constituencies are not compelled to follow orders of a "boss". They have their own needs, wants, and agendas.

There is no doubt that the data captured by smart cities can improve our lives. Yet we know knowledgeable and innovative people who live in smart homes but never use the smart features. These people understand that if they live in a smart city, they are not the customers, they are the product. They choose to interact with people who are not just smart, but kind, generous, and constant. Why should they choose to interact with machines that are simply "smart"?

Smart city *infrastructure* is innovative. Whether *residents* of a smart city can be innovative is another question entirely. Indeed, it is ridiculous to think of entrepreneurs brainstorming a proprietary business plan while Alexa is listening in! Or that new venture companies, traditionally started in garages with scratch funding, would want to locate where they must pay high taxes for smart city infrastructure.

The Global Innovation Forum panel's statements confirmed that we do not want to leave all smart city decisions to corporations and to the profit motive. City officials should become more knowledgeable about the technology, instead of outsourcing all decisions to an IT provider. Decision makers may understand that we need innovation for sustainability, and that this kind of innovation is the common interest of technopolis and smart cities. Smart City 2.0 will need to be designed bottom-up, based on dialogs with many constituencies, rather than top-down with functions dictated by a central authority. Here, the role of different 'actors' in the smart city system also becomes important, particularly citizens, civic society and even media, to feedback information to smart city developers the needs of the community.

We cannot expect the profit motive to yield socially and environmentally optimal results. Yet, authoritarian government measures are distasteful to many, and the opposite government approach—democratically responding to diverse and messy constituent needs—is time-consuming, perhaps too much so in the face of a 2030 Sustainable Development Goals (SDG) deadline.

Creativity and Retro Longing for the Analog World

The State University of New York operates an Innovation Lab in Incheon, Korea. Much of the lab consists of a re-creation of a house that recalls the childhoods of today's Korean knowledge workers. It has wooden walls, a woodstove, stacks of LPs above an old-fashioned radio and phonograph, and a loft where an employee or student may "hide", doubly insulated from the digital environments in the other corners of the innovation lab.

People know that Internet of Things (IoT) devices both deliver and collect data. Knowing that you are watched inhibits your behavior. Inhibition is an enemy of creativity.

If the lab's analog, restful surroundings generate a good idea, you may travel a few meters to the lab's modern section, using digital whiteboards, a recording studio, or teleconferencing rooms to develop the idea further.

Efficiency versus innovation in the smart city, and top-down versus bottom-up design and governance are two very important issues explored in these chapters.

The Chapters

Our authors appear to admire smart city successes in Singapore and Sweden, with Taiwan and mainland China close behind. Their chapters offer rich international perspectives, with authors representing Finland, Singapore, Taiwan, USA, Japan, Germany, South Korea, Italy, Brazil, and with another American author who has been working in Sweden.

Part 1 offers overviews.

(1) Kelechi A. Kalu (USA) draws on his childhood experiences in Nigeria to propose a framework for more humane artificial intelligence in the running of smart cities.
(2) Seo-Jeong Lee, Eung-Hyun Lee and Deog-Seong Oh discuss goal setting and strategic planning for the smart city. They consider the over-arching goals to be efficient city management, improved quality of life, strengthened urban competitiveness, and sustainability.
(3) Ilkka Kakko (Finland) emphasizes the roles of networks and serendipity in dealing with volatility, uncertainty, complexity and ambiguity (VUCA) times. With special reference to research parks and the entrepreneurial start-up scene, Kakko uses networks and serendipity to update the triple-helix and project management paradigms.

Part 2 concerns strategies for smart city innovation.

(4) Ever fond of inventive acronyms, Arcot Desai Narasimhalu (Singapore) explains how the community elements represented by the seven letters "GARLICs" can network to contribute to smart cities' success. The reader must proceed in order to learn what his "CEREAL" acronym stand for. Desai's ideas build on the Brussels Smart City Wheel, which is sharply reminiscent of Gibson's[4] technopolis wheel.
(5) Chien-Yuan Lin (Taiwan) discusses implementation strategies for the promotion of smart city development by both government and private sectors in Taiwan. He describes Taipei's leading and prize-winning measures based on Taiwan's "Digital Nation & Innovative Economic Development Program (DIGI+) 2017–2025".

[4] D.V. Gibson, G. Kozmetsky, and R. Smilor (eds.) (1992). *The Technopolis Phenomenon: Smart Cities, Fast Systems, Global Networks*. Rowman & Littlefield.

(6) Jun Yamashita (Japan) credits a book by Gibson, Kozmetsky and Smilor[4] for originating the smart city term. This 1992 book predated the World Wide Web, yet Yamashita finds that later research defines the smart city solely in terms of the Internet and IoT. This important chapter explores the feasibility of resident-driven innovation in smart city services in Japan. Its findings could explain the early failures of Songdo, which were due to top-down design.
(7) Raimund Bröchler (Luxembourg) and Mathias Seifert (Germany) explain how the science/technology park concept can be extended to a framework of Sustainable Smart City.
(8) Jaehoon Rhee (South Korea) applies the platform approach to Korean technoparks, after recapping the history of Korean industrial policy that led to the technoparks' creation. He explains how the parks shifted from an authoritarian to an egalitarian regime.
(9) Noting the paucity of data on smart city implementation successes, Jun Yamashita returns in Chapters 2 to 7 to place four Japanese initiatives in a planning-cycle context, and to document their stages of success in terms of the societal challenges set forth by Fernandez-Anez *et al.*[5] He echoes Desai Narasimhalu in emphasizing the role of smart city platforms.
(10) Stefano Bresciani, Alberto Ferraris and Gabriele Santoro (Italy) agree with Yamashita that smart cities are not just about making urban environments more efficient and convenient; city officials must face the "societal challenges" that arise from fast-changing economic, technological, and environmental developments. They must put administrative measures in place to address these. Environmental degradation, along with the downward price pressures resulting from globalized economic competition, creates resource shortages. Though some of these shortages can be remedied by price- and technology-moderated substitution, Bresciani and co-authors argue that they heighten the relevance of the Resource-Based View of strategic (public) management.

---

[5] Fernandez-Anez, V., Fernández-Güell, J.M., and Giffinger, R. (2018). Smart city implementation and discourses: An integrated conceptual model. The case of Vienna, *Cities*, 78, 4–16.

(11) With Ben Matheson (USA), Urusha Thapa (Nepal), Y.J. Choi (Korea) and J. Byun (Korea), Fred Phillips (USA) asks why metro-area innovation systems do not have comprehensive information support. The authors claim that benefits of an information system for this purpose would be substantial. Obstacles include competition for philanthropic support among non-profit facilitating organizations, and the general problem of private incentives versus public benefit. Summarizing the experiences of a few prominent Regional Innovation Systems (RISs), the authors suggest a way forward that is consonant with each RIS's maturity level.

Part 3 offers worldwide accounts of smart city best practices.

(12) Sung-Kwang Yang (South Korea) describes how South Korea's Daedeok Research Complex was converted into an innovation cluster in order to develop and productize the devices that drive Smart Home, Smart Factory, and Ubiquitous City, the latter meaning that all digitized services are available to citizens 24 hours per day.

(13) Sang-Ryong Cha (Japan) reminds us that the aim of urban design has been and remains quality of life (QOL). Smart cities will bring efficiency, but efficiency must not degrade QOL, and the pursuit of QOL should not waste resources inefficiently. He goes on to illustrate this balance, with a focus on the energy sector.

(14) Sunghee Lee, Jongkon Kim, and Youngchul Kim (South Korea) graft the smart city concept to "urban regeneration", that is, the revitalization of cities for greater equity, conviviality, and economic competitiveness. They illustrate the hybrid concept with case studies from the Netherlands and South Korea.

(15) Daniel Esashika (Brazil), Gilmar Masiero (Brazil), and Yohann Mauger (USA) reveal that the fastest-urbanizing area of the world is Latin America. In Brazil alone, 26 million people living in urban areas do not have access to water, and 83 million are not connected to sewage systems. These facts present a different set of problems than for richer cities' smart initiatives. Reinforcing that smart cities are not only about IoT, these authors note urban Brazilians' "low physical and social connectivity" and paucity of healthcare

resources, in addition to the low levels of digitalization. They document political barriers to improvement, and relate promising Brazilian developments.

(16) In 2008, one of us (Phillips) had to evacuate from his home inside the city limits of San Diego, due to a wildfire heading in his direction. As drought seasons grow longer and wildfires in California and other western states grow more numerous and difficult to control, urban areas are endangered, and firefighters are in short supply. Sheridan Tatsuno's (USA) chapter illustrates the use of virtual reality and artificial reality (VR/AR, or together, XR) for training candidate firefighters, and more generally for urban disaster preparation. The technologies' potentialities are considerable.

(17) In the final chapter, Enrico Ivaldi and Andrea Ciacci (Italy) develop a method for ranking cities' performance on sustainable services, environment, transport, safety, and innovative economy. They apply the method to twenty Italian cities, drawing conclusions about spatial patterns of success.

The chapters address or imply other vital questions as well:

- How can smart cities support and encourage human-to-human interaction? One of the GIF panelists noted the high expense of providing automated monitors to guard the elderly. How could smart city services help the elderly to guard over each other?
- Is it too much to hope that multiple IT providers—presumably private companies with intellectual property to protect—will allow their systems to talk to each other? If not, could there be other avenues for mutual learning and adaptation across cities served by different providers?
- How can the all-seeing IoT eye reconcile with European legislation on the "right to be forgotten" (the General Data Protection Regulation, or GDPR)?
- What analog amenities support our analog psychology—the psychology that gives rise to our creativity—and what digital amenities support the innovative expression of that creativity?

- Our authors note that urbanization increases apace worldwide. Yet the coronavirus pandemic has spurred those who can, to abandon the cities. Will the smart city be smart enough to retain the mobile educated and affluent?

The editors hope that research on these questions, including the research reported in this book's chapters, will help create cities that foster human growth, creativity, innovation and joy.

Columnist George Will writes[6] that new capabilities for analyzing baseball statistics have led, for the first time, to more strikeouts than on-base hits in the typical game, with base-stealing all but a lost art. He fears this makes baseball less entertaining. He means that spectators do not want to see optimized baseball; instead, we go to games to see outstanding examples of human heart, acuity, and guts. His column might well have been titled *The Downside of Big Data*. Let us hope that the new baseball is not a leading indicator of our cities' futures.

The editors thank Eung-Hyun Lee and Lan Anh Alice Nguyen for editorial assistance, and Series Editor Dr. Tugrul Daim for inviting the volume.

Deog-Seong Oh, Daejeon
Fred Young Phillips, Albuquerque
Avvari V Mohan, Kuala Lumpur
January, 2023

[6]Will, G.F. (2021). *Opinion: Baseball is Losing its Entertainment Value. It's Time to Change the Rules*. URL: https://www.washingtonpost.com/opinions/2021/07/09/time-has-come-save-baseball-by-changing-rules/.

Part 1

Overviews

© 2023 World Scientific Publishing Company
https://doi.org/10.1142/9789811257186_0001

CHAPTER 1

Policies and Rules of Engagement in Traditional and Smart Cities[1]

Kelechi A. Kalu

Political Science and Founding Vice Provost,
International Affairs
University of California Riverside, Riverside, USA
Email: KaluK@ucr.edu

1.1 Introduction

Despite challenges growing up in a war-ravaged region of Nigeria in the early 1970s, I had an exciting childhood because I grew up in a village where everyone knew everyone else. Knowing each other enabled social, economic, political, and cultural collaborations. Indeed, without the close-knit communities, it is doubtful that many of us would have survived the Nigerian civil war in my part of the country. As young people, we went everywhere together—to the stream, to fetch firewood in the bush, walking together to elementary school, or simply gathering in a circle to listen to the elders tell scary and sometimes uplifting stories about a world and experiences we had little to no knowledge about. It also meant that sad experiences were shared because we always walked and worked in groups. As I would later find out when I moved away to Lagos, Nigeria's capital city at the time, that was the social nature of the village.

[1] Divya Joshi provided research assistance on this chapter, which was presented at the 2019 Global Innovation Forum in Daejeon, South Korea. I am grateful to Tomi Daniel for reading the draft of this chapter and providing suggestions that improved this essay.

Moving to Lagos at the end of the civil war was confusing for me because even though people lived together, there was a lack of the same sense of community I had experienced in the village. In Lagos, people spoke differently and in different languages, worshiped differently, and often lacked the collaborative ethics I grew up with in the village, where everyone in the community participated in some way in cleaning their environment or coming to the aid of a community member who needed help. Indeed, the idea of distrust of the intent of those who were different either because of their language or cultures became a common social reality that I had to begin to learn to understand.

Unlike the village, Lagos and the other cities of my childhood in Nigeria, such as Umuahia, Onitsha, Ibadan, and Kaduna had more people, different vehicles, and infrastructures with basic manufacturing and assembling factories. Even at that time, the cities were becoming congested with plastic and petrochemical wastes. And while, over the years, manufacturing facilities have decreased with increased importation of goods from other countries, the congestion and pollution have not abated. The road networks, health services, energy supplies, and other infrastructures in these cities remain undersupplied, and at times hopelessly inefficient, with negative consequences to peoples' health and security. These cities have become quite challenging as spaces for a wholesome life and living in, as pollution from wastes dumped from industrialized countries, poor air quality, dangerous levels of flooding resulting from climate change—partly triggered by the industrial outputs of more advanced countries—continue to deflate the quality of life for the people who live in those cities.

On the other side of the world, advanced cities like New York, Washington DC, Denver, Los Angeles, Columbus, London, Seoul, Beijing, Shanghai, Mexico City and several other cities that I have had the opportunity to have either live in or travel to for work, are not necessarily havens for human health and security either. In many of these cities, overpopulation, homelessness, poor air quality, traffic congestion, high levels of obesity, diabetes, and other health challenges are constant daily experiences needing improvement. Environmental scientists keep reminding us that human actions are largely responsible for changes to the climate, which pose a challenge to all governments, individually and collectively, to find solutions that will enable us to live wholesome lives. With climate change, while some areas are experiencing deplorable levels of flooding, many citizens across the world are unable to find potable water to quench their thirst. What the foregoing scenarios suggest is that contemporary cities are fraught with challenges that require a rethinking of our roles and responsibilities to the planet that has sheltered us for millennia. And it appears that current technological advancements in Artificial Intelligence (AI) are rising to the call for careful collective regional coordination for sustainable solutions to global challenges. While some contemporary cities already have some aspects of AI in their infrastructure or way of life, efforts to transition contemporary cities to full-fledged smart cities that run largely on AI—software and hardware—will challenge everything we have come to know and believe are the domain of human beings. I am interested in exploring the extent to which, as self-interested contemporary humans, we are likely to agree to use AI to work with people or organizations that have control and power over codes and algorithms to change our lives by implementing collaborative and/or coordinated policies with minimal challenges to our common humanity.

1.2 Smart Cities Run on Artificial Intelligence: The Good and the Questions

1.2.1 *The good*

The characteristics of smart cities are already with us in our contemporary life and living. As Jamie Susskind notes, smart cities are "…embedded with dense networks of sensors that allow authorities to keep track of a dazzling array of variables, from noise, temperature, light, air toxicity, and utility usage, to the movement, location, and activity of vehicles and people" [Susskind, 2018]. Software powered by codes and algorithms already activate devices that track our movements, moods, locations, and speech across countries like China and Israel. Some of these devices are also tools used by intelligence agencies in democratic and non-democratic societies to gather information about their citizens. For example, drones are capable of tracking a person in a crowded location until they find the exact individual. Some drones also are deployed to scan and provide data on the conditions of food crops in large farms and monitor plant diseases. Some AI already have the capacity to save lives because they can process data faster than humans and therefore can make more accurate, precise, and objective decisions. Using precise codes and algorithms, such machine learning can save more lives in surgery and reduce human errors. Ongoing research is enabling AI to mine data to provide accurate clinical trials and therefore prescriptions that target diseases on an individual basis. Such targeted and precise decision-making has the potential for reducing the devastating effects of cancer, HIV/AIDS, and other diseases. Machine learning's ability to use early detection techniques is helping prevent health problems such as diabetes, high blood pressure, and stroke before they become emergency medical conditions. Given ongoing successes in these areas, the expectation is that Smart Cities run by AI will also be able to solve problems that contemporary cities have been unable to solve.

In the area of human thinking, software like Grammarly already can help writers complete sentences or make corrections that seem to indicate the software's near ability to read human thoughts. In that respect, advanced AI should have the capacity to read human thoughts, which will be good for, perhaps, solving crimes. At the same time, it is possible that such capacity might intrude on human's most private of spaces—our minds. Already, Siri gives directions, plays music, responds to questions, and even engages in small talks (albeit, currently limited) with humans. Google sings *Happy Birthday* in the morning, turns lights on and off, or dims it on command—actions that eerily mimic human companionship. And then there is Alexa with her ability to respond, prompt, and offer a versatile range of assistance to her human owners. All these are good and convenient technologies for increasing human comfort.

There are enormous reasons to welcome AI in societies. Perhaps these robots will be able to clean our oceans of the suffocating plastics on ocean floors and beaches, saving not just human, but marine lives as well. This also suggests that since AI software is coded to be indifferent to race and nationality, then robot-powered algorithms have the potential to deracialize justice and reduce encounters that lead to the death of certain racial and identity categories, especially in multiracial cities. Perhaps AI can significantly reduce boredom and, eventually, inequality in the workplace by taking charge of repetitive and back-breaking routine jobs like garbage removals, keeping streets clean, and digging trenches.

As we speak, shopping in grocery stores has less and less human interaction experiences as checkout clerks are slowly being replaced with time-saving self-checkout options. Many customer call centers are now staffed by chatbots. In short, whatever can be automated is installed and in use. This also suggests that by using AI, smart cities can be designed to enhance human interactions to improve quality of life, especially for individuals who may have challenges with hearing, speech impediments, and other physical and emotional challenges. AI can predict natural disasters, for instance, storms, earthquakes, wildfires, and deploy non-human machines to minimize the impact of such disasters on humans, especially in the case of wildfires. With AI, people are likely to live much longer. At the current rate of AI's success in different aspects of human life, the challenge will be the impact of that success on the earth's carrying capacity. However, that is a different story. For now, we are happy with life-saving sensors.

In agriculture, AI is already impacting food production by using sensors to indicate when parcels of field crops should be watered, when a segment of a crop farm is infected and what needs to be done to reduce the spread of plant diseases. Simply put, AI technology's ability to use millions of sensors to scan, understand, and act in ways human intelligence is not yet able to do is both effective and impressive. The ongoing research in AI to Augmented Intelligence within the framework of 5G will change everything with its processing speed as it analyzes data faster than humans can comprehend.

On the level of the ability to work with machines, AI processes satellite images much faster than humans, deploys and retrieves data from drones, and advanced AIs may be able to engage in drone warfare between countries—with devastating impact for less industrialized states whose technological capacities are not up to date.

Since these things are now possible, the biggest challenge the world will have to deal with is the fact that AI cannot reason, show emotions consistent with a humane touch, or independently raise capital to invest and produce new constructions. It cannot independently form teams to collaborate for universal good without humans coding and writing the algorithms necessary for AI—advanced or basic—to function. And, this is what worries me about AI—the fact that their marching orders have to be given by people—humans with interests and ideologies that do not always mean well for other humans! Consequently, I am concerned at the speed with which innovation in AI technology is moving without community conversation about how, when, and under what circumstances AI should or should not impact life and living.

1.2.2 *The questions*

With advanced machine learning, what are the likely impacts of AI on our current understandings of our community? How will AI impact governance and collective decision making that have been the staple of human collaboration? What are the likely effects of AI on human relations—especially in the contexts of choices that individuals make in the privacy of their homes that others may not approve of?

At the same time, how might this co-existence with machines affect human employment in intense data reading tasks like X-rays, mortgage processing, corporate tax processing, and stock trading? I am not concerned that humans will be displaced from their jobs. Rather, my concern here is how those humans who control codes and algorithms are likely to deploy AI for non-sustainable communal purposes, like intentionally defrauding the public treasury without community's recourse to recovering their resources because the code and algorithm controller or tech firm may be the only one who knows how and when to deploy the code.

Given the rate at which current technological development and machine learning are advancing across industrialized states, how likely is it that military decisions will begin to rely solely on AI because its predictions will be more accurate and "efficient"? When such a reality is normalized, what then becomes the role of Heads of Government or the relevance of universal treaties that still attempt to regulate conduct at war? To what extent will wars become merely data mining and calculations, with the final decisions for war based on which State has the tools to win without regard to their targets or the fallout on non-targets? How will machine-assisted decision-making impact human conditions when those with genocidal intents are either in control of (or deployed for) the levers of government or simply in charge of codes and algorithms? These are not rhetorical questions. One only has to think of historical atrocities such as Hiroshima, Nagasaki, Armenia, the Holocaust, Indigenous people in the Americas, and Australia, Biafra, Bosnia-Herzegovina, and Rwanda! Humans did say once before—never again!—only to have those with the power to change current algorithms or access to them shy away from the responsibilities of that resolve and decide to do nothing when later victims were different or posed inconvenient choices. With code- and algorithm-powered AI, the opportunity now exists for such neglect and irresponsibility without direct attribution—the irresponsible and convenient argument will be that the AI tool did it!

Perhaps, as important as the convenience that AI is likely to provide by enhancing lives and living is its impact on people's freedom and autonomy in making choices or decisions that transcend their immediate communities. How important will freedom of choice and ethical decisions be in a society where decisions are made by machines that have no real human capacity for empathy, except what is coded into it? In this sense, if algorithms and coding only reflect the data coded into the system, AI is unlikely to know or think about the authenticity of the data coded into the system. To that extent, how much trust and faith should people have in high-tech companies and governments whose interests are not always in alignment with the interests of the masses? What about with machine-based human resources (HR) decisions where implicit bias can be a double-edged sword that can both aid or prevent discrimination in hiring and career advancement processes, depending on what end the codes and algorithms are programmed to pursue? Yes, autonomous driving vehicles (ADV) are likely to be safer than humans, *ceteris paribus*, but what happens in an uncertain situation that calls for a humane and ethical decision on the road? What if the passenger would like to go faster than ADV is programmed to allow? Whose choice and decision take precedence—human or machine? And, to the extent that machine learning allows AI to make decisions for humans, is that not ultimately transforming our existence where humans will now be taught by machines? If so, what effect will such a reality have on our values of individual freedom, especially in democratic societies? The age-old philosophical question of "to be or not to be?" becomes "to learn or not to learn?" In traditional societies, the consequences of 'not to be' were largely minimally individualized. For example, how will machine teaching change the dynamic of an individual's willingness to be different from their society and impact our sense of community? And to what extent will the facial recognition software [Smith and Browne, 2019], which is currently evident in some parts of the world, affect the civil liberties of citizens of both that society and those coming from other societies or countries? And if laws against the use of facial recognition are in place in some parts of the world, does that also mean that the intelligence community does not have the software and the power to deploy such technology without the citizens of such countries necessarily knowing about it? Similarly, if AI is purely in the hands of individuals in a governance-enabling community, how can the community ensure that regulation of innovative inventions result in what the developers were permitted to produce and not beyond it? However, given the intrusive nature of technology for decades, if AI in the hands of individuals can be regulated to avoid disruption of civil liberties and rights, how likely should citizens trust that governments and transnational groups/corporations are not likely to deploy such technology to the disadvantage of the citizens at a convenient time? Finally, what happens if the software that enables the ADV to make decisions is hacked by an unfriendly entity? These questions are not trivial to the extent that humans continue to demonstrate capacities for inhumane decisions on issues such as migration, refugees, genocide, human and drug trafficking, and environmental change actions, objectification and racial hatred of people who are different from us, especially those that we live with in the same cities.

1.3 Politics Process Model Can Undergird Rules in Smart Cities

Under normal circumstances, cities are characterized by some form of communal rule or the other, and decisions are often negotiated political interactions between people whose resources and positions in society afford them the opportunity and/or privilege to negotiate for others. Thus, as in my village community where elders guided collective decisions, in contemporary States and cities, governments make and enforce decisions to guide human and corporate behaviors in their communities in ways that often preserve some public good. Without territorial boundaries—thanks to globalization—smart cities that run on codes and algorithms will pose a particularly difficult challenge to make and enforce laws against faceless coders and writers of algorithms that power AI. While in contemporary societies and States, policies pose benefits to some and losses to others, existing inequities in contemporary States will pale in comparison to the potential inequities that smart cities, without prior rules that guide the conduct and behavior of the owners of high-tech firms and software engineers, will evoke. Thus, how the benefit of smart cities will advance the common good rather than replacing humans and/or creating unbearable levels of inequality in societies across the world deserves our attention.

A key part of what makes us human has been our ability to coordinate our actions, bargain, and come to a consensus (most of the time) on the best way to live together. Hence, the traditional strategy of selecting an issue or problem for the attention of policymakers that often reflects a process of political bargaining, compromise, and consensus will be challenged by the codes and algorithms that undergird smart cities. Consequently, how likely will agenda issues as the outcome of interactions between actors with different resources and interests in conflict remain the bases for reaching commons decisions? In other words, will politics as we know it survive codes and algorithms?

In *Political Science in Theory and Practice*, Ruth Lane argues that the outcomes of issue-based battles are directly related to "the power of strategically situated individuals to understand, confront, and change circumstances" [Lane, 1997]. The contextual framework could be an agenda-setting or decision-making arena where, for instance, a labor union organization would like to see worker-friendly policy against the deployment of AI in the factory but conflicts with companies that prefer robots to human workers. In a contemporary setting, with expert knowledge on both sides, winning will reflect the difference between the resources available to each group. Analytically, the following diagram illustrates the directional outcome of an issue with a contested benefit.

Individual capacity to change existing rules and institutions is largely dictated by the individual or group resources matched against those of the opposition in the context of politics [Kalu, 2004]. As illustrated in Figure 1.1, the outcome of such a political process is either to maintain rules and institutions which support existing structures and people in power, or to change the rule and therefore, people also. However, in Smart Cities powered by codes and algorithms, it is not certain who the actual decision-makers will be—humans or machines. Powered by faceless machines and real humans with control of the codes and algorithm, will the outcomes of contested issues not be predetermined by existing control that tech firms have over innovation and use of information technologies? The outcome will reflect the knowledge that the winners have of the rules, their resources, and the weaknesses of the opposition in each issue or policy. My contention here is that the inherent inequity humans face against others, for example, tech firms, with control of data processing power and capacity for fast-paced research that can be marshaled against humans is likely to create a dangerous level of inequality,

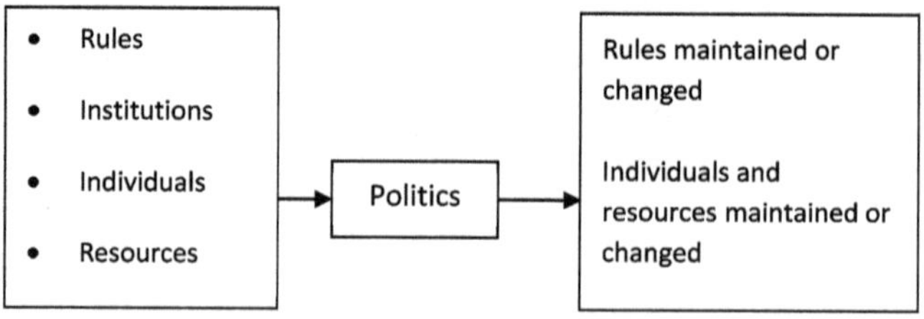

Figure 1.1. Logic of analysis of the politics model.
Source: Adapted from Ruth Lane [1997, p. 64].

if not completely change our contemporary understanding of what it means to live life as participating members of a community.

Existence of a political event (e.g., hacking an election) or a policy issue (e.g., protecting workers' rights) with a different outcome for the participants should require a full description of individual and organizational participants and the resources they bring to the interaction table—control of codes and algorithmic knowledge, economic wealth, cognitive abilities, and information and/or knowledge of the rules—these are what makes community interactions sustainable. When the rules change, it is usually because some groups can maintain or change the rules based on equally accessible resources. With AI, it is not only the rule but also the structure of the game that changes. Those with power over codes and algorithms, therefore, have more than their fair share of the definitions and knowledge of norms, customs, rules, and institutions within which the interaction takes place and will control community resources beyond the capacity of regular people to exist with dignity. Relevant here is understanding whether the rules allow for a single or a long-lasting decision with a long-term impact on the individual participants and the larger society. Knowledge of the rules is important as most political bargaining processes are about controlling the rules and therefore, the values, norms, and processes of maintaining and changing how society is structured and governed [Lane, 1997, pp. 8–10].[2] Thus, with face recognition software, algorithms that can identify and interpret emotions, having the capacity as people to make autonomous decisions will be compromised. GPS already changed how we drive or pay attention to the road. It has already changed our capacity to remember geographic landmarks because the software remembers for us. While this is good, it is the implication for privacy, human dignity, and respect that should be of concern to us.

And as "the process of politics involves the pursuit of interdependent outcomes in the context of participants' beliefs about the importance and nature of just relationships, procedures, and outcomes" [Lamborn, 1997], codes and algorithm power holders will also have the

[2] These characteristics are derived from Ruth Lane [1997, pp. 8–10].

power to redefine these values and remove human vulnerability inherent in interdependent relationships. Then, politics as the strategic interaction of actors with different beliefs, attitudes, values, and goals will cease *ipso facto* as a process that involves solving common problems as public policy. Consequently, for conflicts over the distribution of resources and general issues of political rules within a given context, politics will become an ongoing human-algorithm relationship in smart cities. Conceptually, political interactions will be restructured by (a) the preferred interests and therefore expected outcomes that tech firms and governments expect, (b) each actor's beliefs about changing or maintaining the existing rules will largely reflect values suggested by algorithms, (c) how strongly each actor prefers its outcome and therefore the risk s/he is willing to take will be dictated by data and (d) how well each actor can link issues that could yield the actor's preference will be decided by codes and algorithms.

As individual participants tend to reflect the values of the community they represent, the preferred interests of a specific group could be the maintenance of existing opportunities, such as in the case of information technology firms. Similarly, the resources such participants are willing to invest in ensuring the existing political rules and institutions become a function of how strongly the status quo is preferred. Intuitively then, "changing the rules of politics changes the incentives for political actors; …changing incentives lead to changes in political behavior; … changing behavior changes political institutions and their significance" [Lane, 1997, p. 78]. This then explains the enhanced political activities of such transnational corporations like Amazon, Facebook, and Microsoft, especially since the end of the Cold War.

Understanding how social issues, especially issues with civil rights policy ramifications, become part of the policy agenda requires knowledge of individual participants, their interests, and beliefs about existing institutions and rules. Such knowledge is essential for two reasons. First, it is the elite in authoritarian and constitutional democracies that translate issues from social problems to the policy agenda. For example, issues about unemployment and inequity are often ignored in both constitutional and non-constitutional forms of government. Second, increasingly in contemporary cities—same if not more so in smart cities—while elections are often convenient times for politicians to campaign on issues that appeal to the masses, e.g., poverty reduction in society and social justice, codes, and algorithms, controlled by tech firms will play an increasing role in who gets elected with consequences on policy outcomes.

And, how likely is it that the tech firm elite in a smart city will care about social justice and poverty alleviation without a crisis? Indeed, to what extent are codes and algorithms likely to inform people of such catastrophic events as famine, drought, flooding, corruption, high infant mortality and AIDS, if such information offends the senses of those whose power control codes and algorithms, and or if such information harms the firm's profit margin? For, irrespective of the type of government, it is the elite that have control of maintaining or changing existing institutions and rules, which in smart cities will be States and tech firms having control over codes and algorithms.

1.4 Concluding Remarks

While Hollywood has the potential to entertain, it sometimes provides deep insights into the human condition through movies. In the move *I, Robot*, based on Issac Asimov's Three Laws of Robotics (1950; 2004) decree that:

(1) A robot may not injure a human being or, through inaction, allow a human being to come to harm.
(2) A robot must obey orders given it by human beings except where such orders would conflict with the First Law.
(3) A robot must protect its existence as long as such protection does not conflict with the First or Second Law [Asimov, 1950; 2004].

However, one of the robots refused to follow the Three Laws of Robotics because it had been coded and equipped with algorithms that allow it to have empathy, and to disobey and even harm humans. The challenge that codes and algorithms raise for humans is real because it threatens to change the way we live and understand life and living. While technology is an enhancement in our lives, policy frameworks that humanize these new devices are lagging behind the rapid rate of innovation and adoption. Indeed, it is not even clear that average citizens have options. The deck has already been stacked by tech firms and governments against citizens. In some cases, citizens themselves fall for the shiny exterior of novelty and convenience without probing deeper some of the risks that the technology they desire poses to their health, wellbeing, security, and freedom. Therefore, I propose two simple approaches that might get the conversation started.

First, given different economic capabilities of States and regions, having a global conversation on a policy framework to bring AI and its owners into existing law will be too lame to be helpful. What is needed are regional conversations framed around what needs to be culturally preserved in each region and, based on that, establish a rule—collectively enforceable—that no tech firm will have permission to conduct business in a region without adherence to the cultural markers identified as sacrosanct by the collective will of the people in that region.

Second, school curricula will need to emphasize the interdisciplinary relationships between the humanities and sciences as a basis for future leaders to know that technological advances need to always be evaluated against threats they pose to the elements of what makes us human. Think about Boeing 733 Super Max, the wonder aircraft that was praised for its advanced technology but resulted in the loss of hundreds of lives because of its ability to render humans powerless against exerting control over it in certain circumstances. The lives lost did not need to be lost. Yet, it was not until after two fatal crashes that its manufacturer and regulators were willing to concede that human lives should not be left to software, codes, and algorithms, because when they fail, the catastrophe hits home on humans—not machines. What we think about and how we relate to our thoughts have consequences for the type of knowledge we produce. The anomie that has been in existence in modern society—separating us in preference for sometimes inanimate objects—is partially responsible for the proliferation of technologies that promise to challenge our human dignity.

To me, the foregoing was captured in the following 1997 Apple Advertisement:

> Here's to the crazy ones. The misfits. The rebels. The troublemakers. The round pegs in the square holes. The ones who see things differently. They're not fond of rules. And they have no respect for the status quo. You can quote them, disagree with them, glorify or vilify them. About the only thing you can't do is ignore them. Because they change things. They push the human race forward. And while some may see them as the crazy ones, we see genius. Because the people who are crazy enough to think they can change the world are the ones who do. [Susskind, 2018, p. 6]

In every generation, there have been such crazy ones and misfits and rebels, troublemakers, and round pegs in square holes to whom we owe our sanity and convenience, and the ability to dare more and achieve more. As the next generation of thinkers and inventors use AI to broaden our horizons and expand the limits of what we can contemplate and innovate for smart cities and other aspects of our lives, may the lessons painfully emphasized by disasters like the Boeing 733 Super Max crashes, unintended casualties of drone attacks, and hackings, and malevolent genetic modifications, among others, be central to the innovation, diffusion, and adoption of new technologies. In institutions of learning, may the next generation of thinkers and inventors be educated sufficiently across the disciplines to marry their humanity with projected technological inventions. And may they live with the reminder that humanity stands to enjoy the greatest gains when AI systems are designed to augment human contributions and tap into the complementarity that smart machines and human intuition can give. When these become our reality, all of us will be able to look forward to living wholesome lives in communities that are enhanced but not threatened, by the technologies that we will dare to build.

References

Asimov, I. (1950). *I Robot*, Gnome Press, New York.

Asimov, I. (2004). *Three Laws of Robotics,* in the *I Robot*, a Movie. Davis Entertainment, Los Angeles, U.S.

Kalu, K.A. (2004). *Agenda Setting and Public Policy*. Chapter 1 and Chapter 4. London: Routledge.

Lamborn, A.C. (1997). Theory and the politics in world politics, *International Studies Quarterly*, *41*(2), 187–214.

Lane, R. (1997). Political Science in Theory and Practice: The 'Politics' Model. New York: M.E. Sharpe.

Smith, B. and C.A. Browne. (2019). *Tools and Weapons: The Promise and the Peril of the Digital Age*. New York: Penguin Press.

Susskind, J. (2018). *Future Politics: Living Together in a World Transformed by Tech*. Oxford, UK: Oxford University Press.

© 2023 World Scientific Publishing Company
https://doi.org/10.1142/9789811257186_0002

CHAPTER 2

The Conceptual Framework and Key Strategies of Smart City Planning for Sustainable Urban Development

Seo-Jeong Lee,[*] Eung-Hyun Lee,[†,§] Deog-Seong Oh[‡]

[*]*Department of Architecture Engineering, Chungnam National University, Daejeon, Republic of Korea*
[†]*Department of Architecture Engineering, Woosong University, Daejeon, Republic of Korea*
[‡]*Woosong university and former president of Chungnam National University, Daejeon, Republic of Korea*
[§]*Corresponding author, Email: chris.eunghyun.lee@wsu.ac.kr*

2.1 Introduction

According to the UN World Urbanization Project [2018], the global urbanization rate, that is, the proportion of the world's population living in cities, continues to increase to 54%, and the urbanization rate is expected to rise to about 66–68% by 2050. In particular, in the case of Korea, the urbanization rate rose from about 30% in the 1950s and 1960s, to 91.8% as of 2020. The population concentration into these cities has caused many urban problems including affecting the growth of cities.

So, what are the main problems emerging in modern social cities, and what should the cities of the future look like? The answer for this can be a sustainable city. The global economy grew around cities due to two industrial revolutions from the 19th century to the early 20th century, and accordingly, cities continued to grow, with increasing concentrations of the populace migrating to them and the development of transportation in cities. The development of transportation has influenced urban development by promoting activities in cities and contributing to overcoming spatial constraints. As a result, the use of automobiles increased rapidly, and the convenience of urban residents due to traffic congestion decreased, and economic damage and environmental pollution continued to increase. In other words, the cities, where most of the population reside, have become a "criminal" contributor to climate change, by being a major carbon source.

Urban growth has affected climate change because cities have inherent limitations. As cities grow, they become internally differentiated and specialized into various functional areas to form an urban internal structure [Kaplan *et al.*, 2014]. As cities grew, urban planning became necessary, and through this, it was necessary to develop land for residential and economic purposes and social activities, and green fields outside the city have been developed to provide space for urban activities including recreational. In addition, in order to do the necessary activities in the city, people living in the city have no choice but to live under the city's ever narrowing spatial constraints. This increase in urban growth and traffic volume has been a major cause of increasing urban problems such as environmental pollution, climate change, noise, increased traffic congestion and accidents, loss of pleasant urban space, and increased energy consumption and carbon emissions.

As an alternative to the limitations of climate change, environmental pollution, energy problems, and resource economy that threaten humanity, the development of a "sustainable city" was proposed [Meadow *et al.*, 2004], and a smart city using information communication technology (ICT) and convergence technology in the 4th Industrial Revolution is being presented as a model. Until now, cities have chosen new developments to respond to urban problem improvement and development needs. Mass production and mass consumption based on the existing industrial revolution laid the foundation for the formation and growth of modern cities, but as mentioned earlier, it caused environmental problems and various socioeconomic problems.

Smart cities are recognized as future growth engines to replace fossil fuel-based existing industries while improving urban competitiveness and quality of life by improving fundamental problems arising from the mass consumption structure of modern cities [Hwang, 2016]. Therefore, it can be said that the smart city plan is to realize a sustainable city through changes in the spatial structure and infrastructure development of cities fused with evolved intelligent information technology. Therefore, this study aims to understand the approach of smart city planning from the perspective of comprehensive urban planning and to understand the correlation between smart city goals and sectors and strategies, while identifying the most important areas and implementation strategies to achieve them.

2.2 Goals and Key Implementation Strategies for Smart City Planning

As mentioned above, smart city planning refers to realizing a low-cost, high-efficiency urban spatial structure by promoting changes in urban activity distribution and spatial structure using intelligent information technology, thereby achieving sustainable urban development. It is an urban planning strategy that innovates cities so that modern cities can efficiently and conveniently carry out various urban activities and new lifelong life through convergence with the 4th Industrial Revolution technology, minimizes environmental problems, and builds a smart city ecosystem. Looking at the main goals of modern cities through the development of smart cities, they can be summarized as "securing efficiency of urban management", "improving quality of life", "strengthening urban competitiveness", and "enhancing sustainability" [Albino *et al.*, 2015; Caragliu *et al.*, 2011; Clarke, 2013; Dameri and Rosenthal-Sabroux, 2014; EIP-SCC, 2014; Komninos *et al.*, 2011; KRIHS, 2016; Park and Yoo, 2017; Toppeta, 2010].

In the short term, through the intelligence of urban infrastructure, urban problems are solved with high efficiency and low cost to secure efficiency in urban management, and the quality of life is improved by increasing citizens' convenience, safety, comfort, information access, and openness. In other words, it can be said to increase the efficiency of urban activities and space management by utilizing currently applicable smart city technology without drastically replacing buildings and infrastructure in existing cities. For example, existing cities have been solving problems by expanding roads or constructing new roads to relieve traffic congestion, but smart cities provide road information in real time to induce drivers to use bypass roads or improve traffic problems by controlling traffic signals according to traffic volume. In other words, it actively utilizes information and communication technology to provide detailed information to urban facility users while increasing the convenience of use so that users can follow a wiser way to increase the capacity of the facility [Lee, 2019].

In the long run, the technology of the 4th Industrial Revolution, urban space, and infrastructure will converge to create a different urban space than before, and will significantly change urban activity patterns. Just as cars have innovatively changed the structure and characteristics of urban spaces in modern society, these future smart city technologies will significantly change the urban space constraints and industrial ecosystems. For example, the traffic environment will change rapidly due to the emergence of new means of transportation such as self-driving cars and drones, while various urban infrastructure related to this will be required, and the city's spatial structure is expected to change to a different shape. Just as the urban space structure and scope centered on walking until the mid-19th century changed to a modern city linked to the development of automobile transportation in the early 20th century, the development of urban space and transportation linked to the technology of the 4th Industrial Revolution will transform the city into a completely different shape.

In order to maximize urban efficiency through smart city services in the short term and lead to innovative changes in infrastructure, spatial structure, and facilities for the application of smart city systems in the long term, sustainable innovation must be promoted. Establishing a new industrial ecosystem promotes the development of smart cities while strengthening urban competitiveness.

It can be said that it aims to build a balanced and sustainable city for environmental, social, and economic development [Yigitcanlar, 2016; Yigitcanlar *et al.*, 2019; Zhao, 2011]. Smart cities are developing in various meanings by country and city, but smart cities generally agree on the concept of a city that integrates physical infrastructure, IT infrastructure, social infrastructure, and business infrastructure into a single framework [Harrison *et al.*, 2010]. Therefore, in this chapter, considering the hierarchy and characteristics of smart city goals and comprehensive urban planning,[1] the main concepts of smart city are divided into three components: (1) urban infrastructure, (2) urban environment necessary to reduce environmental load and (3) urban services that meet citizens' needs and respond to industrial structure changes.

First is the aspect of urban infrastructure. Smart cities can be defined as cities that make networks and services of traditional cities more efficient by applying digital and information and communication technologies (European commission). By using advanced information and communication technology to comprehensively monitor information on the main situations of major infrastructure (road, rail, water, power, buildings, etc.), it achieves efficient maintenance and management of cities, optimization of urban resources, and maximization of service supply [Hall, 2000]. Therefore, smart cities will establish an intelligent urban infrastructure and mobility system that systematically

---

[1] Hwang [2016] explains that smart cities have major differences in terms of 'urban structure, function, and service'. As a major configuration aspect of smart cities, Shehadi *et al.* [2015] is divided into Digital Enrolling Levers, Next-Gen Infrastructure, Ecosystem Components, and Smart Service, while Wang [2016] and Yang Sung-min [2017] are divided into urban spaces, urban environment, urban life, and policies. Based on the configuration aspects proposed in previous studies, this study resets the 'smartization, service, platformization, and eco-friendly' (Ministry of Trade, Industry and Energy, 2016) into three smart cities (urban infrastructure, urban environment, and urban service) applicable to the urban space hierarchy.

and intelligently combines physical infrastructure through urban planning and design and intelligent information technology (AI, IoT, Big Data, sensors, etc.) to support them, and establish a low-cost, high-efficiency, and low-energy urban spatial structure.

Second is the aspect of the urban environment. Smart cities are defined as cities that enable energy conservation, efficient allocation of resources, and improvement of environmental quality, including major environmental issues (energy, resources, and environment) in digital cities [Nam and Pardo, 2011; Zygiaris, 2013]. In order to implement an efficient energy-sharing city, it means switching from a linear resource consumption structure to a circular resource consumption and recycling structure through systems such as smart grid-based energy systems and ICT-based resource circulation systems, and creating a pleasant green environment for real-time monitoring.

Third is the aspect of urban service. Recently, smart cities are paying attention to the lives and social and economic values of citizens that change through the application of smart technology [Caragliu *et al.*, 2011; Cretu, 2012; Eger, 2009]. Smart cities should address social and economic needs, participatory governance, efficient management of resources, and economic value creation. To this end, smart cities can provide open platform-based living and policy services for citizens and improve the quality of services based on voluntary cooperation and participation of citizens. In addition, it is necessary to foster citizens or businesses and provide a new business environment where innovation can be continuously created through the opening of various urban data so that a smart innovation ecosystem can be created. This is not just a problem limited to the growth of the urban economy, but a key role in promoting smart city development by providing low-cost, high-efficiency facilities, infrastructure, and services through technology-based innovation.

Therefore, the main strategies presented to realize the concept of a smart city centered on the four goals pursued by smart city planning can be expressed as shown in Figure 2.1 by summarizing the strategies presented from the perspective of smart city infrastructure, smart city environment, and smart city service.

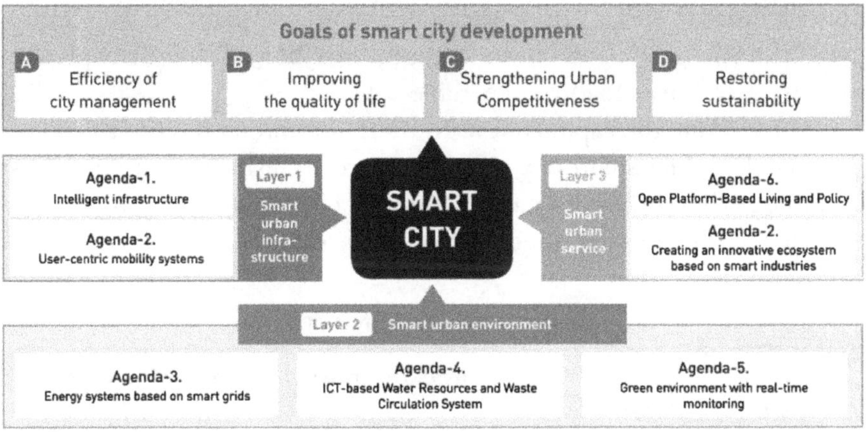

Figure 2.1. Goals and key implementation strategies for smart city planning.

2.3 Smart Urban Development Case Study

2.3.1 *Overview*

Among the cases of promoting smart cities as an urban planning model for realizing a sustainable city using intelligent information technology, we find cases that meet the following three criteria and examine the main contents of smart city planning.

1. Cities that establish and promote urban plans, including the concepts and goals of smart cities,
2. Three aspects of the smart seat framework set up in this study, a city that comprehensively deals with detailed plans for seven core sectors, and
3. A city that promotes specific plans and designs, systems, and detailed projects.

Based on these three criteria, three case areas were selected: (1) Amsterdam Smart City Plan (Netherlands), (2) Stockholm Royal Seaport Development project (Sweden) and (3) Busan Eco-Delta City project (Korea), and major strategies for smart city planning were examined. The background and basic status information on smart city plans in the case area are as follows (see Table 2.1).

Amsterdam, the capital of the Netherlands, promoted the European Union's (EU) first smart city with the aim of transforming it into a sustainable, efficient, and a carbon-neutral city in response to climate change. Amsterdam is encouraging different organizations to work

Table 2.1. Case overview.

Project	Amsterdam	Stockholm Royal Seaport	Busan Eco-Delta City

Area	219 km^2	2.4 km^2	2.5 km^2
Population	About 820,000 people	About 20,000 people	About 20,000 people
Objectives	• An efficient city based on high technology • Sustainable city • Climate Change Neutral City • 40% reduction in CO_2 emissions and 25% reduction in energy consumption compared to 1990 by 2025	• Improvement of health, comfort, safety and life quality • Ensuring ecological, social and environmental sustainability • Reduce CO_2 emissions by 1.5 tons per population by 2020 • Zero use of fossil fuels by 2030	• Building a sustainable society • Reduce greenhouse gas emissions • Renewable energy plus 20% and recycle 100% • Data-based smart city operation • Improving the quality of life • Citizen-led governance innovation • Create an innovative ecosystem and create jobs

together on innovative projects to create a living city where people can live and work happily. It can be said that it is a private-led smart city plan implemented to solve urban problems, centered on the Amsterdam Smart City (ASC) consultative body consisting of citizens, research institutes, global companies, and government officials. It is encouraging public–private–citizen cooperation by providing an innovative platform that can propose and accept ideas online and offline that can solve social issues (discovering new industries, creating jobs, etc.). Currently, more than 100 projects are underway with six themes: (1) Digital City, (2) Energy, (3) Mobility, (4) Circular City, (5) Governance and Education and (6) Citizen and Living.

The Royal Seaport Project in Stockholm, Sweden, launched a smart city project for the regeneration of industrial areas in 2008 based on sustainable development. Royal Seaport has established a smart city plan aimed at improving health, comfort, safety, and quality of life, promotes urban development, and aims to achieve zero fossil fuel use by 2030 with a specific goal. The total area of Royal Seaport, which was redeveloped in the port area located in the Baltic Sea, is 2.4 square kilometers and consists of houses, work spaces, and commercial spaces. With the aim of completion by 2030, it is striving to establish an ICT-based sustainable urban system by establishing a specialized network involving consultants, suppliers, and system suppliers. Various business plans are being established in a total of six areas: (1) Energy, (2) Traffic and Transportation, (3) Architecture, (4) Water, (5) Waste and (6) Landscape Planning.

The Eco-Delta City (EDC) Project in Busan, Korea, is a new city development project, with an area of 2.2 square kilometers and a planned population of 8,500, designated as a national pilot project to convert into a smart city based on the existing Ubiquitous City (U-City) development plan. It aims to be completed in 2023, and is being promoted as a smart city pilot project that integrates major urban functions such as housing, commerce, and research and development (R&D) with the support of the government and applies various advanced technologies on a trial basis. The Busan EDC aims to create a sustainable city that can improve the quality of life socially, environmentally, and economically by solving urban problems by reflecting and participating in civic requirements. The Busan EDC aims to become a "global innovative growth city where nature, people, and technology meet to accelerate future life", and introduces the "fourth industrial revolution technology" to set all citizens as key values to "improve the quality of citizens' lives", "balance opportunities", and "inclusive growth". As a practical strategy for this, it is planning three specialized strategies—(1) Smart Tech City, (2) Smart Water City and (3) Smart Digital City—and seven core contents in consideration of Busan's urban problems and core values.

2.3.2 *The application status of smart city development strategies*

2.3.2.1 *Smart urban infrastructure*

Intelligent infrastructure

The Intelligent Infrastructure strategy plays a role in integrating urban information collected by linking an efficient urban spatial structure with advanced information and communication infrastructure and laying the foundation for smart city services optimized.

First, in terms of "improving sustainability", an efficient compression and complex urban space that can secure the self-sufficiency function of the city and increase spatial connectivity is envisioned. This can be realized through the application of urban planning techniques represented by distributed intensive development, transit oriented development (TOD) development, and complex use development. It can be said to be an effective plan to rationalize land use, minimize unnecessary traffic generation, and secure a pleasant open space through high-density development centered on urban development and public transportation axes (BRT, etc.) (for Amsterdam, Royal Seaport, and Busan EDC). In addition, by creating a complex district where various urban functions such as housing, business, commerce, and culture are integrated, it was intended to realize the proximity to work (for Royal Seaport and Busan EDC).

Second, in terms of "securing efficiency of urban management", it is to establish an intelligent advanced information and communication infrastructure to collect and share urban information in real time. It combines urban spaces with core technologies such as broadband information and communication networks, information collection infrastructure (IoT, CCTV, sensor, RFID, QR code, near field communication (NFC), etc.), data hub (Big Data, cloud, etc.), and security system (blockchain, etc.) to enable real-time collection, sharing and utilization of various data in cities (for Amsterdam, Royal Seaport, and Busan EDC).[2]

Third, in terms of "securing efficiency of urban management", an integrated management system of urban data is established to appropriately respond to real-time changes and utilize data by monitoring cities in an integrated manner. Efforts were being made to build multiple databases such as 3D virtual spaces (3D maps, digital twins, etc.) in urban spaces. In addition, it was used to provide convergence services (urban disasters, energy, transportation, etc.) required by citizens and to establish, apply, operate, and manage urban plans (for Amsterdam and Busan EDC). In particular, the Busan EDC establishes an integrated urban operation center to collect data urban spatial information, and manages and analyzes it in real-time based on AI.

Smart mobility system

The Smart Mobility System refers to a system for providing safe, convenient, and environmentally sustainable urban transportation services by integrally controlling and responding to traffic information in real time based on advanced smart transportation systems and road networks.

First, it is necessary to establish a green transportation-oriented road network that can secure eco-friendliness and comfort in terms of "improving sustainability" and "improving quality of life". Road networks are planned by securing the connection of green transportation (walk, bicycle, public transportation, etc.), and public transportation transfer centers are established at major transportation nodules to increase the convenience of using green transportation. In addition, a physical environment is established so that pedestrians and bicycle users can move comfortably and safely, and institutional traffic regulations

---

[2] In particular, in the case of the Busan EDC, the 'Data Market' platform was operated to enable private companies and citizens to create, trade and share data needed.

(telecommunications priority, speed limit areas, etc.) are provided (for Amsterdam, Royal Seaport, and Busan EDC).

Second, it collects and manages real-time traffic information in terms of "securing efficiency of urban management", and establishes and operates a smart transportation integrated platform that can increase safety and convenience by providing customized transportation services in terms of "improving quality of life". In all cases, an integrated transportation operation center was established to monitor, collect, and share real-time traffic information based on traffic information collection infrastructure (GPS, CCTV, IoT, etc.). The collected traffic information was identified and provided real-time traffic information sharing system, public traffic information system (BIS, ITS, VMS, etc.), real-time traffic signal control system, automatic traffic control, and parking information system (for Amsterdam, Royal Seaport and Busan EDC).

Third, in terms of "improving sustainability", it will promote the introduction of eco-friendly future new transportation with high energy efficiency and low carbon emissions. It will activate the introduction of transportation (trams, electric/hydrogen vehicles, electric bicycles, etc.) using eco-friendly energy such as electricity and hydrogen, and build charging infrastructure. It will introduce new public transportation (BRT, LRT, PRT, etc.) and promote plans to upgrade road systems for this. Royal Seaport and Busan EDC will also actively provide car-sharing services (car sharing, P2P, etc.) with high citizens' convenience and growth potential (for Amsterdam, Royal Seaport and Busan EDC).

2.3.2.2 *Smart urban environment*

Smart Grid-based energy system

The Smart Grid-based Energy System integrates the energy supply and demand of smart cities based on advanced ICT to convert the energy system to an energy independent city and respond appropriately to climate change (see Table 2.2).

First, with the aim of "improving sustainability", the government plans to convert to a new and renewable energy system to minimize carbon emissions and supply energy that can reduce fine dust. Through this, in terms of "strengthening urban competitiveness", economic

Table 2.2. Analysis of application status in smart urban infrastructure.

	Application status			Goals of smart city planning				Relationship between main strategies and goals
Strategy	**Amsterdam**	**Royal Seaport**	**Busan EDC**	**A**	**B**	**C**	**D**	
Intelligent infra-structure	• Development of distributed and concentrated system • TOD	• Mixed-use development • TOD	• Mixed-use development • TOD	○	○		●	Efficient, compact and mixed-used urban space → Reduce carbon emissions, reduce energy consumption, and increase urban self-sufficiency
	• Digital infrastructure	• ICT infrastructure • Smart grid	• IoT, 5G, etc. • Urban data hub • Security system	●	○			Building an intelligent advanced information and communication infrastructure → Realization of urban control based on advanced technology

(Continued)

Table 2.2. (*Continued*)

Strategy	Application status			Goals of smart city planning				Relationship between main strategies and goals
	Amsterdam	Royal Seaport	Busan EDC	A	B	C	D	
	• Collecting and managing urban information and in real time • Urban 3D map		• Establishment of an urban data integration center • Building a virtual space (digital twin, etc.)	●				Integrated management system of urban spatial data → Increase the efficiency of real-time management in cities
Smart mobility system	• Connectivity between bicycle transportation methods • Establishment of a comfortable and safe bicycle environment	• Green transportation network (walk, bicycle and public transportation) • Establishment of a comfortable and safe walking and bicycle environment	• Green transportation network (walk, bicycle and public transportation) • Creation of a pleasant environment for walking		●		●	Road network centered on green transportation. → Provide comfort and safety, and reduce carbon emissions

				A	B	C	D	
	• Integrated Transportation Operation Center • Customized transportation service provided	• Intelligent Transportation Systems	• Building the next generation intelligent transportation service (C-ITS) • Collecting and sharing real-time traffic information	●	●			Building an integrated transportation platform → Increasing the management efficiency of urban transportation → It provides safety and convenience
	• Electric vehicles and charging infrastructure • Car sharing system	• Introduction of new and renewable energy transportation • Car sharing system	• Electric vehicles and charging infrastructure • Car sharing system		○		●	Activation of new transportation → Reduce carbon emissions and increase energy efficiency

Note: (A) Securing efficiency of urban management, (B) improving quality of life, (C) strengthening urban competitiveness and (D) enhancing sustainability.

effects can be expected through fostering new industries related to renewable energy. Considering the characteristics of the city, renewable energy (solar heat, solar power, geothermal power, wind power, hydrothermal, hydrogen, biogas, etc.) production and storage facilities were planned and utilized as small distributed power plants such as complexes and buildings (for Amsterdam, Royal Seaport and Busan EDC).

Second, for the purpose of "improving sustainability", high energy efficiency performance of buildings and urban facilities to reduce energy consumption is secured. In order to realize this, a passive design for optimizing building energy is basically adopted (for Amsterdam and Royal Seaport). In addition, detailed projects are being promoted to apply smart technology-based energy monitoring systems (BEMS, HEMS, all three cases) and automate the control of building facilities (shading, shielding, lighting, heating, security equipment, etc.) (for Amsterdam, Busan EDC). In addition, some are promoting the automation of control of public facilities such as streetlights (for Amsterdam).

Third, for the purpose of "securing efficiency of urban management", an integrated urban energy management system will be established to systematically control, monitor, and respond real-time to energy supply and demand, and maximize energy efficiency for the purpose of "improving sustainability". Through this, in terms of "strengthening urban competitiveness", economic effects can be expected through fostering new industries related to energy two-way transactions. It operates an integrated urban energy management center based on smart grid and energy storage technology. Information collected through energy control facilities (Smart Grid, AMI, ESS, etc.) throughout the city can be monitored, and the entire process of energy "consumption forecast-supply-demand response" can be systematically managed. Through this, two-way energy transaction services led by citizens as prosumers[3] (excessive energy transaction systems between individuals, renewable energy transaction systems, etc.) are activated (for Amsterdam, Royal Seaport and Busan EDC).

---

[3] A Prosumer is an individual who both consumes and produces—a portmanteau of the words products and consumers.

ICT-based resource circulation system

Smart cities maintain, restore and continuously operate and manage the resource circulation system as much as possible through the establishment of an ICT-based resource circulation system that considers the connection and synergy between resources (water resources, waste, etc.) at the urban level.

First, for the purpose of "securing efficiency of urban management" and "improving sustainability", an integrated resource management model and system for minimizing resource and energy waste and reducing carbon emissions will be introduced at the urban level. Focusing on the Resource Integrated Management Center, the total amount and movement through the overall circulation process of water and waste resources are monitored in real time (for Busan EDC). In particular, in the case of Royal Seaport, an Eco-Cycle 2.0 model was planned as an integrated management system to explain important associations and synergies between the flows of energy and resources (i.e., water and materials). The Eco-Cycle 2.0 model introduced ICT as a tool to connect systems and systems composed of energy, water and materials, and through this, we tried to solve the energy and carbon emission problems (for Royal Seaport).

Second, with the aim of "enhancing sustainability", a sustainable water resource management system with smart technology will be established to enable microclimate control, prevention of water disasters (drought, flood, etc.), and create a pleasant environment. Through ICT-based water management specialized technology, clean water can be supplied to homes by remotely monitoring the entire water supply process such as water quality, water quantity, and leakage in real time. In the case of the Busan EDC rainfall, eco-friendly excellent reuse systems such as low-impact techniques (LIT), eco-filtering, and eco-friendly reservoirs are established and reused. Small-scale smart water purification facilities are distributed in the city center of Amsterdam, Royal Seaport and Busan EDC, and sewage treatment water that has undergone an altitude treatment process is supplied to various necessary water (maintenance water, washing water, landscaping water, etc.) (for Royal Seaport and Busan EDC).

Third, smart cities monitor the entire process of waste treatment in terms of "securing efficiency of urban management" and promote the advancement of waste treatment systems to efficiently use waste and save potential energy in terms of "improving sustainability". Based on advanced waste management and treatment facilities, waste collection, classification, selection, and treatment are automated, and real-time monitoring (waste status, for Busan EDC) is also linked to recycling and energization (biological fuel, HP system, local heating, etc.) and citizens can check the waste treatment process (for Amsterdam, Royal Seaport and Busan EDC). Through this, smart cities can reasonably reduce waste and increase recycling rates.

Green environment with real-time monitoring

The Green Environment, which enables real-time monitoring, uses cutting-edge technology to create a pleasant urban environment, and maintain and manage it in real time to respond to climate change and provide space for ecosystem experiences and educational activities for citizens (see Table 2.3).

First, in terms of urban planning, an integrated green and blue network will be planned to reduce carbon emissions with the aim of "improving the quality of life" and "improving sustainability". It preserves ecosystem diversity by expanding sufficient high-quality green and water spaces in the city and creating ecological spaces (biotops, etc.). Green and blue networks will be established by reorganizing green and water space plans so that point green areas (building, etc.), shipping green areas (connected green areas, buffer green areas, etc.), green areas (parks, etc.), and water spaces can be connected. In addition, the Green-Blue Network allows walking access anywhere in the city, enabling nature-friendly daily life and relaxation (for Royal Seaport and Busan EDC).

Second, with the aim of "securing efficiency of urban management" and "improving sustainability", a park green space management system that combines advanced smart technologies will be established to increase the efficiency of maintenance and management of park green areas. For the created green and ecological spaces, information inquiry and utilization are facilitated through dataization, and integrated management of green and blue networks in the city is promoted. It

Table 2.3. Analysis of application status in smart urban environment.

Strategy	Application status: Amsterdam	Application status: Royal Seaport	Application status: Busan EDC	Goals of smart city planning: A	B	C	D	Relationship between main strategies and goals
Smart grid-based energy system	• Establishment of new and renewable energy generation facilities	• Renewable energy generation facilities	• Establishment of small distributed power generation facilities for new and renewable energy • Integrated operation and management of urban distributed power generation facilities	○	○	●	●	Renewable energy activation system → Reduce carbon emissions, reduce fine dust → Fostering new industries
	• Passive design. • Building Energy Monitoring System (BEMS, HEMS, etc.) • Automation of lighting equipment	• Passive, active, energy plus house plans	• Zero Energy House • Building Energy Monitoring System (BEMS, HEMS, etc.) • Automation of construction equipment control • Smart streetlights	○	○		●	Securing high energy efficiency performance of buildings and facilities → Reduce energy consumption and optimize energy efficiency

(*Continued*)

Table 2.3. *(Continued)*

Strategy	Application status			Goals of smart city planning				Relationship between main strategies and goals
	Amsterdam	Royal Seaport	Busan EDC	A	B	C	D	
	• Smart grid • Establishment of an integrated energy management center • Two-way energy trading service	• Smart grid • Two-way energy trading service	• Smart grid • Establishment of an integrated energy management center • Two-way energy trading service	●		●	●	Establishment of an integrated urban energy management system → Increase efficiency in urban energy management → Increase energy efficiency → Fostering new industries
ICT-based resource circulation system		• Establishment of an integrated management model for the resource circulation system	• Monitoring and managing the resource circulation system	●			●	Introduction of an integrated resource management model and system → Increase efficiency of urban resource management → Reduce carbon emissions and increase energy efficiency

	• Eco-friendly rainwater reuse system	• Eco-friendly rainwater reuse system • Advanced sewage treatment facility	• Automated water quality, water quality, and leak management facilities • Rainwater reuse system •Advanced sewage treatment facility	○	○	●	Building a water resource management system → Regulating microclimate, preventing water disasters, and increasing efficiency of using water resources
	• Waste recycling and energization facilities	• Automation system for waste collection, classification, and disposal • Tracking the waste disposal process and monitoring in real time • Waste recycling and energy recycling facilities	• Automation system for waste collection, classification, and disposal • Tracking the waste disposal process and monitoring in real time • Waste recycling and energy recycling facilities	●	○	●	Advanced waste disposal system → Increased efficiency of urban waste management → Reduce carbon emissions, minimize waste, and increase resource efficiency

(Continued)

Table 2.3. *(Continued)*

Strategy	Application status			Goals of smart city planning				Relationship between main strategies and goals
	Amsterdam	Royal Seaport	Busan EDC	A	B	C	D	
Green environment with real-time monitoring		• Preservation and expansion of high-quality green areas and water spaces • Preserving ecological diversity • Green and blue network plan	• Preservation and expansion of high-quality green areas and water spaces • Preserving ecological diversity • Green and blue network plan		●		●	Integrated Green and Blue Network Plan → Increase the comfort of the city → Control the microclimate, reduce carbon emissions
	• Intelligent park facilities, trees, and ecological monitoring system		• Nature experience and field training using VR/AR	●	○		●	Establishment of an eco-friendly and ecological space management system that combines advanced technologies → Increase the efficiency of green environment management → Reduce carbon emissions

	• Sensing the environment and collecting environment data through IoT • Monitoring environmental data and responding in real time • Provide information on the city's environment	• The city's environmental information service	• Sensing the environment and collecting environment data through IoT • Monitoring environmental data and responding in real time • Provide information on the city's environment	○	●		●	Monitoring the urban environment (atmosphere, temperature, CO_2, noise, etc.) → Increased access to information → Reduce carbon emissions

Note: (A) Securing efficiency of urban management, (B) improving quality of life, (C) strengthening urban competitiveness and (D) enhancing sustainability.

efficiently maintains and manages pleasant eco-friendly spaces, facilities, and ecosystems (such as trees) through environmental sensing and combination with IoT technologies. In addition, it is used as a space where eco-friendly and ecological spaces can be experienced through VR and AR, and educational activities can take place (for Busan EDC).

Third, it is the monitoring of urban environmental information (air, temperature, carbon dioxide, noise, etc.) to disclose urban environmental information to respond appropriately and induce changes in citizens' eco-friendly education and behavior in terms of "improving sustainability". Through environmental sensing and IoT technology, air (fine dust, temperature, humidity, carbon dioxide, etc.), water quality, soil pollution, and noise are continuously monitored as environmental data in urban spaces. In addition, by analyzing and predicting this, local customized measures are established in real time, information is disclosed, and eco-friendly behavioral changes of citizens are induced (for Amsterdam and Busan EDC).[4]

2.3.2.3 *Smart urban services*

Open platform-based living and policies

The "Open Platform-based Living and Policy" section is promoted to provide citizens with knowledge and information collected and produced in the city based on the open platform, and to allow citizens to directly participate and contribute to create various services necessary for smart cities.

First, it provides customized information services and online and offline smart education to enhance citizens' access to information and improve the quality of education with the aim of "improving the quality of life". Citizens' sensory information (i.e., weather, transportation, community information, etc.) collected and produced based on advanced information and communication technology was provided to citizens in real time. Citizen education required in the future society can be

---

[4] Advanced the fine dust analysis system in the city with fine dust monitoring equipment: "fine dust measurement—transmission of information to digital twins—simulation of the installation location of fine dust reduction devices on a city scale—installation of fine dust reduction devices at the same location as the analysis results—actual results".

operated through continuous provision of educational content using virtual reality (VR) or augmented reality (AR), Massive Open Online Courses (MOOCs), and remote education. In addition, workshops or fairs on smart cities are held so that you can experience smart technology and discuss urban issues (for Amsterdam and Busan EDC).

Second, with the aim of "improving the quality of life", a smart care (i.e., health, crime prevention, safety, disaster prevention, etc.) system will be established to maintain the health and safety of citizens and to "secure the efficiency of urban management". Based on the design and arrangement of smart care facilities in major areas of life (i.e., health, criminal, safety, and disaster), we intend to establish an integrated control center and provide services. The integrated control center supports citizens to conveniently access and use services through various devices (i.e., wearable devices, smartphones, kiosks, etc.) in various areas of the city, such as life information, health, crime prevention, safety, and disaster prevention collected through CCTV, GPS, IoT (for Busan EDC).

Third, with the aim of "improving the quality of life", citizen-participating online and offline platforms such as the Living Lab will be established to increase the participation of citizens in policies. Based on open government portals and services, it can be realized by sharing government administrative data (real status of municipal government activities, policies, etc.) online. In addition, through private-led online and offline platforms (Living Lab, etc.), citizens can freely propose ideas about policies, social problems, projects, and services necessary for cities using the knowledge and information provided (for Amsterdam,[5] Royal Seaport and Busan EDC).[6]

Creating an innovation ecosystem based on the smart city industry

The "Smart city industry-based Innovation Ecosystem Creation" section is required to establish an innovative ecosystem to create a new economic

---

[5] Citizens can propose and operate various projects on the online platform ASC web page, or experience smart city projects and exchange ideas in the exhibition space of the offline platform Amsterdam Smart City Experience Lab.

[6] The online website will invite citizens' ideas, and offline booths will provide exhibition contents such as smart city promotional videos, brochures, and VR, thereby making Busan EDC a citizen-participating design city.

foundation for smart cities, continuously strengthen the foundation for knowledge, information, and technology commercialization (see Table 2.4).

First, with the aim of "strengthening urban competitiveness", it will strengthen the capabilities of related institutions (company, research institute, etc.) in the region and create an environment to foster R&D centered on new smart city-related industries. Research and production districts of smart city-related knowledge convergence industries are included in the main contents of urban planning, and R&D-related support policies (policy investment, education programs, etc.) and services induce the revitalization of the 4th industry-based convergence industry. In particular, through the introduction of regulatory sandboxes, legal and institutional support such as regulatory special cases for the introduction of new technologies and services, support for verification of new technologies, and permission for pilot projects is reflected in the plan (for Royal Seaport and Busan EDC).

Second, with the aim of "strengthening urban competitiveness", startup companies will be attracted around the startup support center and various activities will be supported to foster high-tech-based startups, jobs, and create profit models. By providing a sharing platform for programs, data, and facilities to attract startup companies, centering on the startup support center, it can support new technology development and the influx of creative talents. Amsterdam and Busan EDC also aims to attract domestic small- and medium-sized enterprises, startups and global companies overseas through mentoring systems linked to large companies, developing investors and partners (for Amsterdam), and operating "Open Task Sandbox"[7] and "Open Factory"[8] (for Busan EDC).

Third, smart city projects are being promoted by establishing a close industry–academic cooperation system to secure sustainable economic growth engines and create innovative results (knowledge, information,

[7] A system that supports startup spaces, core infrastructure, and investment funds for a certain period of time through programs to foster and support innovative companies and gives companies an opportunity to participate in the creation of Busan EDC Smart City.

[8] As a new concept factory to support innovative companies, it supports experimental and production equipment such as digital devices, software, and 3D printers to help companies participate in open tech sandboxes freely produce various products.

Table 2.4. Analysis of application status in smart urban services.

Strategy	Application status			Goals of smart city planning				Relationship between main strategies and goals
	Amsterdam	Royal Seaport	Busan EDC	A	B	C	D	
Life and policies based on open platforms	• Customized information (shopping, finding directions, etc.) for citizens. • Continuous civic education (remote education, etc.) • Hosted by online and offline workshops and fairs		• Provision of information on urban life that citizens can feel • Expand education using smart technology		●			Citizen-customized information service and on-offline smart education → Provide quality education and increase access to information
		• Community activity support center	• Integrated smart care control system • Data collection, sharing, and integrated management • Real-time monitoring and remote response to emergencies	●	●			Establishing a smart care (health/criminal/safety/disaster) system → Increased efficiency of urban management → Safety and health

(*Continued*)

Table 2.4. *(Continued)*

Strategy	Application status			Goals of smart city planning				Relationship between main strategies and goals
	Amsterdam	Royal Seaport	Busan EDC	A	B	C	D	
	• Citizen-participating online and offline platforms (Living Lab, etc.)	• Citizen-participating online and offline platforms (Living Lab, etc.)	• Citizen-participating online and offline platforms (Living Lab, etc.)	○	●			Establishing an open online and offline platform for citizen participation → Increased access to information → Participation in civic policy
SmartIndustry-based Innovation Ecosystem		• Research and production district development of knowledge convergence industry • Support policies related to Research and Business Development (R&BD) and innovation	• R&D cluster development of knowledge convergence industry • Support policies related to R&BD and innovation • Regulatory sandbox			●	○	Fostering R&BD in the new smart city industry → Strengthen regional smart industry capabilities and secure economic growth engines

	• Start-up support Center • Start-up support and fostering service		• Start-up facility • Sharing platform for programs, data, and facilities	●	○	Support for start-up activities → Expanding smart services, jobs creation, and fostering future industries
	• Academia-Industry Cooperation Center • Exchange of policy, technology, and information	• Facilitation of R&BD based on academia-industry cooperation • Support for networking and collaboration between companies	• Collaboration support facilities and services • Exchange of policy, technology, and information	●	○	Establishment of an academia-industry cooperation system → Creation of innovation outcomes and secure economic growth engines

Note: (A) Securing efficiency of urban management, (B) improving quality of life, (C) strengthening urban competitiveness and (D) enhancing sustainability.

technology, etc.) with the aim of "strengthening urban competitiveness". It establishes an industry–academic cooperation center, establishes an industry–academic cooperation system using an open process (open-sharing-collaboration) capable of public–private–citizen cooperation (for Amsterdam and Royal Seaport), and supports technology development and commercialization by utilizing remote collaboration services of universities-business-research institutions (for Amsterdam). In addition, policies, technologies, and information are exchanged through memorandum of understanding (MOUs), international forums, and open seminars, and cooperation among participants is strengthened (for Amsterdam and Royal Seaport).

2.4 Concluding Remarks

Now, our city is threatening the lives of citizens due to the urban concentration of the population and the limitations of the resource economy, threatening the sustainability of the city. The urban population is increasing beyond the size of the city's acceptable population, while the lives of urban residents continue to deteriorate due to housing shortages, traffic congestion, excessive energy consumption, and environmental pollution, and has a serious impact on the global environment.

Since the Second Industrial Revolution, cities have been pursuing development-oriented strategies to improve problems occurring in cities based on technological development. In order to overcome the lack of urban infrastructure and the poor urban environment due to the increase in population, the city continued to expand with the development of transportation. Due to the limitations of growth in the existing method due to low growth in social and environmental problems caused by supporting mass production and mass consumption, it has become difficult to expect sustainable urban implementation with the current urban planning and development method. The 4th Industrial Revolution is a way to realize a sustainable city, and it is said that a smart city model fused with intelligent information technology is a solution to humanity's problems such as urban problems and environmental pollution. Therefore, smart urban planning is to convert urban spaces into low-cost, high-efficiency cities through space development and old space improvement using ICT, and efficient management and operation of urban spaces and facilities. From this point of view, to understand the core strategies and their application in smart city planning, we looked at urban planning projects being promoted by three aspects—urban infrastructure, urban environment, and urban service—and analyzed smart city goals.

As a result, there are four main goals of urban development in smart city planning: (1) securing efficiency of urban management, (2) improving quality of life, (3) strengthening urban competitiveness and (4) improving sustainability; and seven major strategies are suggested in terms of smart city infrastructure, smart city environment, and smart city service. The main points to be considered through the projects being promoted for each strategy are summarized as follows.

First of all, in order to achieve the goal of "securing efficiency of urban management", it is necessary to apply strategies to integrated control, monitoring, and real-time response based on high technology, provide high-quality smart services, and increase maintenance and management efficiency of each city sector. In order to achieve the goal of "securing efficiency of urban management", the application of strategies in the "intelligent infrastructure" and "smart mobility" sectors in the smart urban environment and strategies in the "smart grid-based energy system" sectors are recognized as major implementation strategies. In particular, strategies related to the establishment of an integrated management system are being treated as important.

Urban strategies related to building a pleasant urban environment, increasing citizens' participation in policies and access to information, securing citizens' health, safety, and convenience, and providing quality education are being promoted to achieve their goals. In particular, it can be seen that it is important to apply strategies in the "green environment capable of real-time monitoring" in terms of smart city environment and the "open platform-based life and policy" in terms of smart city services.

In order to achieve the goal of "strengthening urban competitiveness", strategies to strengthen corporate capabilities in the region, secure economic growth engines, create innovative performance (knowledge, information, technology, etc.), and foster jobs are adopted as major strategies. Among them, the application of strategies in the "Smart Industry-based Innovation Ecosystem" section in terms of smart city services is the most important, and the economic effects of the new energy industry can be expected through the application of the "Smart Grid-based Energy System" section in terms of smart city environment.

In order to "improve sustainability", detailed goals are set and strategies are adopted to reduce carbon emissions, reduce energy use, control microclimate and prevent disasters, recycle and efficient use of resources, reduce fossil fuels and convert to renewable energy. In order to achieve the goals of "improving sustainability", strategies in the "smart mobility system" in terms of smart urban infrastructure, "smart grid-based energy system", "ICT-based resource circulation system", "green environment capable of real-time monitoring", and "smart industry-based innovation ecosystem creation" in terms of smart city services are important. This is because smart services and urban activities are the most active sectors with the development ICT, and it is the most efficient way to promote smart city implementation in the near future by expanding smart service and urban management based on the establishment of a smart industry ecosystem.

According to the main characteristics of each case, Amsterdam and Royal City Port's smart city plans set specific energy and carbon reduction goals, promote smart city plans from an environmental perspective, and expect carbon reduction, urban environment improvement, and low energy consumption. On the other hand, the Busan Eco Delta City case focuses on the goal of strengthening urban competitiveness and adopts the establishment of a smart industry-based innovative industry ecosystem as a key strategy.

However, the current smart city development plan is evenly applied to general urban planning, smart city infrastructure construction, and smart city operation and management in terms of urban infrastructure, but systematic strategy establishment is insufficient. In addition, the linkage between the strategies proposed in the smart city plan is not clear, and each strategy is presented like an independent strategy, suggesting the risk that the smart city (development) plan can only be recognized as converging ICT into urban spaces, facilities, and infrastructure. Therefore, smart city planning can be said to be a new city model to realize a sustainable city by improving many urban problems that existing cities have while achieving new goals of cities using various 4th Industrial Revolution technologies.

Cities were basically physical spaces. Cities were spatial concepts that supplied and consumed space and resources to enable various activities of urban residents, and moved and communicated for active networking. From this point of view, many people think of creating a new state-of-the-art city armed with cutting-edge ICT in high-rise buildings. However, the goal of a smart city is to create a sustainable city in the future by fundamentally innovating solutions to the many urban problems that modern cities have—fine dust and environmental pollution, crime, waste and resource circulation, conflict between classes, traffic jams and parking, and energy overconsumption. Therefore, it is recognized as the only alternative to realizing a sustainable city by building a smart city ecosystem as a new urban growth engine while improving urban problems through the application of core strategies of smart cities. Currently, many smart city development projects are being promoted around the world, and efforts are required to continuously materialize the original smart city development model suitable for each city by sharing and mutual feedback on the experiences of global advanced cases in the future.

References

Albino, V., Berardi, U., and Dangelico, R. (2015). Smart cities: Definitions, dimensions, performance, and initiatives, *Journal of Urban Technology*, 22(1), 3–21.

Byun, M., Kim, M.-H., Yi, C. and Park, M. (2018). *Social Issues in the Smart City Era and Tasks of the Seoul Metropolitan Government*. Seoul Research Institute.

Caragliu, A., Bo, B., and Nijkamp, P. (2011). Smart cities in Europe, *Journal of Urban Technology*, 2(18), 65–82. DOI: 10.1080/10630732.2011.601117.

Clarke, R.Y. (2013). Smart cities and the internet of everything: The foundation for delivering next-generation citizen services, *IDC Government Insights*, October 2013. URL: https://www.cisco.com/c/dam/en_us/solutions/industries/docs/scc/ioe_citizen_svcs_white_paper_idc_2013.pdf.

Cretu, G.L. (2012). Smart cities design using event-driven paradigm and semantic web, *Informatica Economica, 16*(4), 57–67.

Dameri, R.P., and Rosenthal-Sabroux, C. (2014). *Smart City: How to Create Public and Economic Value with High Technology in Urban Space*. London: Springer.

David, H.K., Steven, R.H., and James, O.W. (2014). *Urban Geography*, 3rd edn. New York: John Wiley & Sons.

Eger, J. (2009). Smart growth, smart cities, and the crisis at the pump: A worldwide phenomenon, *I-WAYS — The Journal of E-Government Policy and Regulation*, *32*, 47–53. DOI: 10.3233/IWA-2009-0164.

European Innovation Partnership on Smart Cities and Communities (EIP-SCC). (2014). *Operational Implementation Plan: First Public Draft*. URL: https://smartcities.at/wp-content/uploads/sites/3/operational-implementation-plan-oip-v2-en2.pdf.

Hall, R.E. (2000). The vision of a smart city, *Proceedings of the 2nd International Life Extension Technology Workshop*. Paris, France.

Harrison, B., Eckman, R., Hamilton, P., Hartswick, J., Kalagnanam, J., Paraszczak, and Williams, P. (2010), Foundations for Smarter Cities, *IBM Journal of Research and Development, 54*(4), 1–16.

Hwang, J.S. (2016) *Smart City Development Prospects and Korea's Competitiveness*. National Information Society Agency: Seoul, Korea. [Korean].

Komninos, N., Schaffers, H., and Pallot, M. (2011). Developing a policy roadmap for smart cities and the future internet, in: *eChallenges e-2011 Conference Proceedings*. IIMC International Information Management Corporation.

Korea Research Institute for Human Settlements (KRIHS). (2016). *A Study on Policy Measures to Strengthen Smart City Competitiveness*. Presidential Commission on Architecture Policy of Korea. [Korean].

Lee, S.I. (2019). *Smart City Planning*. Communication Books, Seoul: Korea.

Meadows, D.H., Randers, J., and Meadows, D.L. (2004). *The Limits to Growth: The 30-year Update*. Chelsea Green Publishing Co.

Nam, T., and Pardo, T.A. (2011). Conceptualizing smart city with dimensions of technology, people, and institutions, *Proceedings of the 12th Annual International Conference on Digital Government Research*, June 12–15, 2011. DG.O 2011, College Park, MD, USA.

Park, J., and Yoo, S. (2017). Critical understanding of current implication of Smart city focusing on information and communication technology, governance, sustainability and urban development, *Korean Association of Space and Environment Research*, *27*(1), 128–155.

Shehadi, R., Khoury, R., Karam, D., and Rahbani, J. (2015). *Smart Cities: A Gateway to Digital Life*. Booz Allen Hamilton.

Toppeta, D. (2010). The Smart City vision: How innovation and ICT can build smart, "liveable", sustainable cities, *Think Report 005/2010*. The Innovation Knowledge Foundation. URL: https://inta-aivn.org/images/cc/Urbanism/background%20documents/Toppeta_Report_005_2010.pdf.

Wang, G. (2016). *A Study on Basic Planning of Zero Energy Smart City*, Korea Research Institute for Human Settlements (KRIHS): Sejong.

World Urbanization Prospect. (2018). *The World's Cities in 2018*. United nations Department of Economic and Social Affairs Population Dynamics. URL: https://population.un.org/wup/Publications/.

Yang, S.-M. (2017). *Planning Methodology and Elements for a Smart City Master Plan*. Unpublished doctoral dissertation, Sungkyunkwan University, Suwon, Korea.

Yigitcanlar, T. (2016). *Technology and The City: System, Applications and Implications*. New York: Routledge.

Yigitcanlar, T., Kamruzzaman, M., Foth, M., Sabatini-Marques, J., Costa, E., and Ioppolo, G. (2019). Can cities become smart without being sustainable A systematic review of the literature, *Sustainable Cities and Society*, *45*, 348–365.

Zhao, J. (2011). *Towards Sustainable Cities in China: Analysis and Assessment of Some Chinese Cities in 2008*. New Work: Springer.

Zygiaris, S. (2013). Smart city reference model: Assisting planners to conceptualize the building of smart city innovation ecosystems, *Journal of the Knowledge Economy*, *4*(2), 217–231.

Websites

Amsterdam Smart City. URL: https://amsterdamsmartcity.com.

Busan Eco-Delta City Project. URL: https://busan.ecodelta-smartcity.kr/.

e-country index of Korea. URL: https://www.index.go.kr/potal/main/EachDtlPageDetail.do?idx_cd=1200.

European Commission. URL: https://ec.europa.eu/info/eu-regional-and-urban-development/topics/cities-and-urban-development/city-initiatives/smart-cities_en.

Stockholm Royal Seaport Project. URL: https://international.stockholm.se/city-development/the-royal-seaport/.

Part 2

Strategies of Smart City Innovation

© 2023 World Scientific Publishing Company
https://doi.org/10.1142/9789811257186_0003

Chapter 3

How to Prepare for the Wicked Problems of the Postnormal Era: Introducing an Evaluation Model for Vucability Improvement

Ilkka Kakko

Karostech Ltd., Sotkuma, Finland
Email: ilkka.kakko@karostech.fi

3.1 The Coronavirus Crisis—A Shocking Introduction to the Postnormal Era

There are several illustrative terms to describe the future we have already entered. Even before the coronavirus crisis (i.e., the COVID-19 pandemic), we saw a slow but steady transformation away from the Normal Era, where circumstances were widely characterized by the Industrial Era, a Knowledge Society and a highly predictable business environment. However, the global pandemic and unforeseen actions that brought entire nations and businesses to a halt in order to control the outbreak and limit the damages in first months of 2020 have really accelerated the transformation. Historically, pandemics have forced humans to break with the past and imagine their world anew. This one is no different. It is a portal, a gateway, between one world and the next; between the Normal Era and the Postnormal Era.

We would like to emphasize some academic perspectives and definitions which were presented before the coronavirus outbreak. 'Postnormal times' is such a definition introduced by Sardar *et al.* [Ravetz, 1993; Sardar, 2010; Sardar, 2015; Sardar and Sweeney, 2016]. Postnormal times are characterized by three 'C's— (1) complexity, (2) chaos and (3) contradictions. Especially interesting in this context is the role of contradictions, as Sardar argues:

> "Organizations are becoming fast-and-loose, reconfiguring around social networks instead of business processes, becoming more decentralized and autonomy increases, more egalitarian. We will belong to our networks–which are our own–and not to institutions that require us to subordinate our interests and selves".[1]

And as the notion of VUCA has gained a lot of interest in recent times, we explain the background here. VUCA is a term originally created within the US Army to describe the volatile (V), uncertain (U), complex (C) and ambiguous (A) times we are facing today.[2] In order to understand and cope in VUCA conditions, Paul Kinsinger, a professor of business intelligence at the Thunderbird School of Global Management in Phoenix, Arizona, and Karen Walch, a professor and consultant at the Thunderbird School of Global Management, have introduced a 'new' leadership paradigm VUCA Prime. It explains the challenges and offers some guidance:

(1) Volatility is mitigated by "vision", a clear-cut master statement of where an organization is headed. When confronted by volatility, leaders need to communicate clearly and make sure their intent is understood.
(2) Uncertainty yields to "understanding", the deliberate ability to "stop, look, and listen". In uncertain situations, leaders need to

[1] Boyd, S. (2013). How the Postnormal Era will Change Everything. URL: http://www.stoweboyd.com/post/49946797759/how-the-postnormal-era-will-change-everything.
[2] URL: https://www.scribd.com/document/440363399/Living-and-Leading-in-a-VUCA-World-Thunderbird-School. Accessed: 8.8 2016.

make sure they get fresh perspectives and remain flexible with regard to solutions.

(3) Complexity is checkmated by "clarity", the deliberate effort to make "sense of the chaos". In complex situations, leaders need to make sure to collaborate with others and stop seeking permanent solutions. To paraphrase an adage, do not let "perfect" become the enemy of "good enough".

(4) Ambiguity is matched by "agility", the ability of a leader to communicate across people and organizations instantly and to move quickly in applying solutions. When confronted by ambiguity, leaders need to listen well, think divergently, and set up incremental dividends. This is captured in the concept of "wirearchy", as opposed to "hierarchy"—where social networks that allow you to engage the insights of many trump the brilliance of any one person.

As seen from this short introduction, VUCA environments, as a consequence of the Postnormal Era, set radically new requirements for business activities in all parties involved. The pressure is felt in the offices of policymakers, in regional development agencies, in business communities, on an individual level, and even in societies and economies. The rule seems to be: "Disrupt—or be disrupted!" In these highly unpredictable conditions, individuals, communities, businesses and economies face an urgent need to create new behaviors and better collaboration patterns. We will examine in this chapter some solutions for survival and success, and recommend measures on how to cope with VUCA conditions.

3.2 The Challenges within Ecosystems and Innovation Intermediaries

Taking into account the hard evidence of the challenges that the emerging turbulent VUCA environment is bringing, it is surprising that the management teams of science and technology parks (STPs) and other innovation intermediaries so far have only limited interest in the benefits gained by the vucability improvement approach we are presenting in this chapter. The conventional Triple Helix model is still a dominant way to organize STPs, even though the limitations of that approach in the emergent VUCA world are so evident. The concept of the Triple Helix of university–industry–government relationships initiated in the 1990s by Etzkowitz [1994] and Etzkowitz and Leydesdorff [1995], encompassing elements of precursor works by Lowe [1982] and Sábato and Mackenzi [1982], and it interprets the shift from a dominating industry–government dyad in the industrial society to a growing triadic relationship between university–industry–government relationships in the modern knowledge societies.

The Triple Helix was dominant for a good 30 years until the Quadruple Helix was introduced by Carayannis and Campbell, and it was a significant improvement to the Triple Helix, because end-users or consumers were added as a fourth element [Carayannis and Campbell, 2009]. However, both the Triple Helix and the Quadruple Helix models still mostly represents the old, industrial push-model with linear pipeline thinking, which works properly in predictable and stable conditions. However, the emergence of Postnormal Era changes the business environment dramatically and these models should be supplemented or even replaced with more flexible, proactive, non-linear and attractive combinations of ecosystem development and platform thinking.

One perspective we want to highlight here is the economic costs of running a Triple Helix model and for some reason, this has not been widely discussed so far. The Triple Helix operations require many highly paid (in many cases by taxpayers' money) officials, managers, and directors acting as gatekeepers and facilitators. The model introduced in the following Section 3.3 has proved to be cheaper and offer more impactful solutions to replace or at least supplement the existing Triple Helix practices. We do not want to underestimate the value of the Triple Helix or Quadruple Helix approaches, knowing that it is not easy to implement the Triple Helix or Quadruple Helix optimally in any real-life contexts of science and technology parks.

The elementary reason that new solutions and especially platforms have lately captured many business leaders' imaginations is that they enable "pull-based" approaches. In the past, sellers have been limited by the economics of scale thinking both in production and distribution, meaning that they simply made an efficient batch size of what they sold and delivered it to the marketplace. This kind of push-based approach is very efficient if business forecasts and strategic foresight is accurate—and can at least be profitable if, failing that, the marketer is able to alter demand with its pricing and advertising. However, when this thinking is applied in times of the Postnormal Era, there are too many and too big "Ifs". Thus, it is obvious that today we need a holistic model of building innovation ecosystems, which takes the increasing of complexity within business landscape and socio-technical systems seriously as a challenge [Rabelo and Bernus, 2015]. Our statement is that the VUCA approach is a relevant starting point to build and utilize platforms, to develop the business ecosystem and STI communications in science and technology parks, in other innovation intermediaries and especially in meta-ecosystems.

To understand the ecosystem thinking and SMEs' multiple roles in that, we have to consider all the participants in the ecosystem. If we want to create value addition to societies, we must understand the original message of David J. Teece [1986; 2006] and take it into serious consideration. He stated that distribution of the profits from innovation is a broader issue than an entrepreneurial issue. Firms, innovators, suppliers, customers and many other stakeholders want profit from innovation. Profiting from innovation is not a simple issue. The traditional Triple Helix stakeholders (government, academy, and industry) are, in the real world, split in many different entities like startups, SMEs, effectual entrepreneurs, anchor companies, freelancers, students, wanna-be entrepreneurs, public organizations, social entrepreneurs, NGOs, associations, researchers, lead users, venture capitalist, and seed funding organizations etc.

Innovations are developed in complex adaptive systems. Often, new ideas and businesses emerge from a combination of these entities and increasingly, when complexity, chaos and contradictions open new windows of opportunities. Typically, SMEs, startups and freelancers are elementary actors in this, but not only as creating business for themselves. The often-neglected parallel roles of SMEs and entrepreneurs in the ecosystem include to be a customer, outsourced expert, resource pool, intermediary, promoter for the entire ecosystem, learning opportunity (for students) and a part of some consortium like a Collaborative

Networked Organization (CNO) and a vital node connecting ecosystems on meta level.

We are sensing that ecosystems will be the decisive entities, when evaluating success factors on the future competitive landscape. Actually, we believe that competition in the future will increasingly happen not anymore between companies, but rather between ecosystems. For most of the local ecosystems, joining meta-ecosystems will be the only way to ensure competitiveness in the global landscape. In fact, even their existence is a matter of belonging to a more viable and competitive meta-ecosystem.

The global startup ecosystem landscape is strongly dominated by the US-based hubs, such as Silicon Valley and Boston. These most successful global ecosystems will keep draining talents and ambitious startups from all over the world and thus become even stronger. Unfortunately, the means to fight against this trend are very limited. By applying the meta-ecosystem approach in other parts of the world, we have a feasible solution to tackle this challenge. The market has proven that coaching and support services need to be combined with top talent and latent resources across the local ecosystems, especially in less developed ones.

Our work in defining meta ecosystems and their viability in VUCA times has created interest in European domain specific ecosystems (see Cases 1 and 2). Our definition as it stands at the moment is as follows:

> "Meta ecosystem is a synergetic and interdisciplinary combination of digitally interconnected domain specific ecosystems, powered by platform thinking and effective digital and physical trust building practises". [Raita, 2020]

3.3 Some Building Blocks for the Survival and Prosperity in the Postnormal Era

There are some proven solutions that help us to survive and to prosper in environments, which are affected by complexity, chaos and contradictions—"the forces that shape and propel postnormal times"

[Sardar, 2010]. Platform thinking, ecosystem development and serendipity management are all either separately, or even more importantly when combined, the solutions to cope with the VUCA environments. In fact, all three of them can become also the underlying power to cause chaos, contradictions and non-linear developments in many industries, and therefore disrupting many traditionally managed businesses. Science and technology park management and even more widely understood—the innovation intermediary business as such—is no exception to that rule. In this chapter, we explain these three elements of vucability from the perspective of STP management and ecosystem operator, and will give some practical guidelines on how to create environments and practices capable to survive and even prosper in VUCA conditions.

3.3.1 *Platform thinking*

At the moment, there is a lot of debate both in business and in academia about the disruptive elements of platform thinking. The recent evidence about the way established platforms and business models have totally transformed entire business areas, such as the hospitality industry (by AirBnB), taxi business (by Uber), music delivery (by Spotify) and publishing (by Facebook) is strong, and this trend is gaining even more momentum, when more industries will be hit by the newcomers introducing powerful applications of platform thinking. We argue that this platform-based innovation thinking will be the most powerful single disruptor of traditional business areas and business models in the near future. This is easily understood after having a closer look at Table 3.1. It shows the benefits gained by platform thinking and it is extremely interesting to see how platform thinking is by definition a perfect solution when VUCA conditions prevail.

We will show some preliminary models on how to create an evaluation and mapping system of "vucability", which we define as the ability to prosper in VUCA conditions. This framework shows the organizational (or individual) capabilities to cope with the VUCA environment and the vucability level can be evaluated using some understandable

Table 3.1. Pipeline-thinking versus platform-thinking.

Pipeline-thinking	Platform-thinking	Main benefits of platform-thinking
Command and control Gate-keepers	Collaboration Low entry, zero-friction	Engagement Creativity
Institutions Shareholder's value Career	Entrepreneurship Stakeholder value Collectives	Sustainable ecosystem thinking
Project management	Serendipity management	Disruption
'From me to you' – lectures Conferences Get and Give Teaching	'From anybody to anybody' – sessions Un-conferences Give and Get Learning	Apprentice attitude, Diversity
Innovation intermediaries Brokering Regional/national approach	Competence platforms, Community building Global approach	New and unique combinations of competences
Push principle Individuals as consumers and resources	Pull principle Individuals as co-creators and supports	Attraction, Gravity, Flexibility More effective allocation of talent (self-organized)

Source: Ilkka Kakko and Kari Mikkelä [2015].

and simple metrics. This evaluation will cover the entire universe of platforms (and the maturity level of ecosystem development and the density of serendipity thinking)—and identifies all types of communication, trading, technological and competence platforms.[3] For the main part, it will be based on the critical elements presented in

[3] For more information about this classification, see Kakko [2017].

Table 3.1. The massive power of platform thinking is well described by Sangeet Paul Choudary [Choudary *et al.*, 2016] in his argument that:

> "Firms that once sought advantage based on the strength of their internal resources and channel access now face competitors that harness armies of connected users and ecosystems of resources.
>
> (1) Platforms will displace high cost gatekeepers with meritocratic crowds;
> (2) Platforms will aggregate disconnected players in fragmented industries; and
> (3) Platforms will unlock new value from spare resources and user-generated content".

Choudary *et al.* [2013] highlights with that argument very clearly, that in order to remake the role that experts play inside your business organizations, you should leverage their user capabilities and build social curation and reputation systems to employ the collective intelligence and judgment of your users. When talking to ecosystem operators while putting together a Meta Ecosystem consortium (as seen in Case 2), we clearly discovered that when this curation and hence collective intelligence is taken to the cross-ecosystem level, we call it the meta-level, the synergy and energy of the total can be leveraged to new spheres.

We want to note that this distinct need can be covered by organizing cross-ecosystem trust building on the meta-ecosystem level and introducing competence platforms for team-building and idea elaboration purposes. The maximum impact is gained by connecting consumers to their best product options, regardless of source, through data-driven matchmaking. The company that builds an OpenTable for consumer finance, considering appetites for risk and reputations of products that deliver on promises, would help buyers make sense of the dizzying array of complex and disconnected products. The value for the company providing the platform will be enormous, not only in terms of sales and profit, but also in understanding the consumer behavior and the fundamental needs on the market.

Finally, you may solve a consumer problem in your industry by marshaling spare resources. For instance, if you are in transportation, build systems that employ other people's trucks before expanding your own fleet.

3.3.2 *Ecosystem development towards meta-ecosystems*

As seen in Table 3.1, the traditional Triple Helix model is designed in a way that it fits perfectly for circumstances where pipeline thinking can be applied. It is very much 'push'-based, highlights the role of brokering and employs in many stages the different type of 'gatekeepers'. In most cases, the Triple Helix solutions are regional initiatives with multiple objectives trying to fulfill the expectations of a variety of stakeholders. Unfortunately, in the turbulent and changing VUCA conditions, the Triple Helix model often is not able to provide the expected results and impact. Many scholars have criticized the concept of the innovation ecosystem to be vague and having no analogy to biological ecosystem. Because of this scientific criticism, we define that innovation ecosystem consists of the actors, material entities, and intangibles. Actors are the members of the academia, government, small business companies, investors and industries, leaders of government investment programs and business intermediators. The intangibles are the complex relationships that effectively promote competitiveness of the whole innovation ecosystem.

Efficiency gains are closely linked to discovery, technology demonstration, product and service development and market demonstration. Inventing and demonstration are two key functions of an innovation ecosystem. This definition of an innovation ecosystem is close to the discussion of the National Science Foundation of the US [Jackson, 2011]. So, another type of innovation intermediary structure is needed, and that is offered by novel ecosystem initiatives based on platform thinking and serendipity management. Although eco-prefix is in many cases unnecessary and even misleading as Oh, Phillips, Park and Lee argued at the 2014 UNESCO-Daejeon Global Innovation Forum,[4] in

[4] For the critics of using eco-prefix in these contexts, see D.S. Oh, F. Phillips, S. Park, and E. Leed. (2016). Innovation ecosystems: A critical examination, *Technovation*, 54, 1–6.

this context, the eco-prefix is genuinely justified because these environments actually are nature-like environments with no 'push' type of management. The evolvement of the system is based on principles of self-organization, bottom-up approach, grassroot level activities, peer-to peer support and the use of a breeding environment type of platform. So, in that respect, they come close to the Virtual Organization Breeding Environments, which were comprehensively researched and developed in the ECOLEAD project [Camarinha-Matos *et al.*, 2008].

The platforms and new business model innovations are an elementary part of modern innovation ecosystem building, because they offer the necessary low-entry, zero-friction access to all participants in the innovation ecosystem. When a well-working competence platform is available, then the vital characteristics of platform thinking like collaboration, community building, 'pull'-principle in the form of the gravity of the platform and 'born global' approach can all be achieved. This combination of new features actually will call for a new entity, which we have named in our research and development work as *a meta ecosystem*.

3.3.3 *Serendipity management as a key factor in VUCA environment*

It is clear that in order to thrive in the VUCA environment, one has to have the ability to benefit from unexpected, chaotic and contradictory circumstances. This is the reason why understanding serendipity is one of the top qualifications in the Postnormal Era. In fact, one of most recent definitions of serendipity is "the art of benefiting from unexpected".[5] It is a useful, simple and yet comprehensive enough description of this expressive word.

In academia, the notion of serendipity is often explained as a process where something unexpected, odd, or contradictory—let it be an event, encounter, accident, test result or teambuilding action—will generate insight, and that insight will lead to action and value creation. Often, a prepared mind or at least certain type of preparedness is also mentioned as a vital prerequisite of the serendipity process.

[5] This definition was created in a panel discussion with serendipity practitioners in the context of "Open Innovation 2.0 Conference" in Espoo, Finland on July 2015.

The definition of serendipity originates in 18th century, when an English nobleman Horace Walpole, who read the Persian fairy tale "The Travels and Adventures of Three Princes of Serendip", was fascinated by the ability of the princes to discover new things. He coined the term serendipity and described the way "the princes were always making discoveries, by accidents and sagacity, of things which they were not in quest of" to his long-term correspondence friend, Horace Mann. Step by step, the notion started to circulate and gain momentum, first among librarians and scientists, then also in business contexts (like pharmacy and chemistry), in such a way that serendipity is at the moment a fashionable 'buzzword'—and in danger to lose its original expression.

For the purposes of this chapter, serendipity can be best explained by using Horace Walpole's original definition, which was further elaborated by Robert K. Merton 1950 [2004, p. 24]: "Serendipity is a quality of mind, which through awareness, sagacity, and good fortune allows one to frequently discover something good while looking for something else" [Merton and Barber, 2004]. This is a comprehensive definition widely accepted also in academia; it highlights the important sagacity perspective and applies as well on the individual and community as organizational levels. So, the widely misused definition of serendipity being 'a lucky accident' is totally inadequate and misleading; if a shortened version is needed, then we can use the notion of 'accidental sagacity', which was also used by Horace Walpole himself.

This leads us to the definition of serendipity management, which we will use in an organizational context: "Serendipity management is a comprehensive set of tools and facilitation methodologies, which by the help of tailored workspace design—both physical and virtual—and through the facilitation of unexpected encounters and collective insight, will support the emergence of new combinations of competences and the generation of breakthrough ideas" [Kakko, 2014]. The revolutionary use of serendipity management as a solution to adapt in the VUCA conditions is explained in Table 3.2, where the characteristics of project management are compared with serendipity management paradigm. We want to emphasize that project management of course is the only way to solve and implement highly predictable challenges where the outcome is clear (i.e., in construction projects etc.), but it is often

Table 3.2. Project management versus serendipity management.

Characteristics	Project management	Serendipity management
Approach	Project	Journey, exploration
Type of innovation	Directional	Intersectional, sustainable
Type of human resources	Homo faber	*Homo ludens, homo aestheticus, homo creatives*
Organization	Fixed in the beginning	Flexible during the process
Focus	Effective process	Best possible result in the end
Structure	Closed innovation	Open innovation
Mission	Goal decided in the beginning	Vision decided in the beginning
Competence search	While defining the project	Training camp approach
Resources, time schedule	Fixed	Flexible
Management style	Command and control	Connectivity and collaboration

Source: Kakko and Inkinen [2009].

unsuitable when the vision is ambitious, and unexpected turnouts and events will affect the implementation process.

3.3.4 *Cyclic innovation model as a framework*

As illustrated in previous chapters, the vital elements for scoping in the VUCA world are platform thinking, ecosystem development towards meta-ecosystems and serendipity management. There is so far no perfect framework for innovation processes, but the Cyclic model of technological entrepreneurship and innovation (CIM) presented by Berkhout *et al.* [2006] and Trott *et al.* 2016 is coming close to that. It does not see innovation processes as a linear pipeline model, instead it illustrates it as an organization model based on cross-disciplinary networks. Our development work has also indicated that in order to improve the CIM framework, we should be talking about an organization model based on ecosystems—or even meta-ecosystems—instead of networks.

The cyclic model of technological entrepreneurship and innovation links startups, SMEs and even corporate entrepreneurship to four domains (see Figure 3.1):

(1) scientific exploration,
(2) technological research,
(3) market transitions and
(4) product creation.

Between scientific exploration and technological research there is the Natural and Life Sciences Cycle. Between scientific and market transitions there is the Social and Behavioral Sciences Cycle. Between technological research and product creation there is the Integrated Engineering Cycle. Finally, between market transitions and product creation there is the Differentiated Service Cycle. These four cycles are important dynamic forces in innovation ecosystems. The Natural

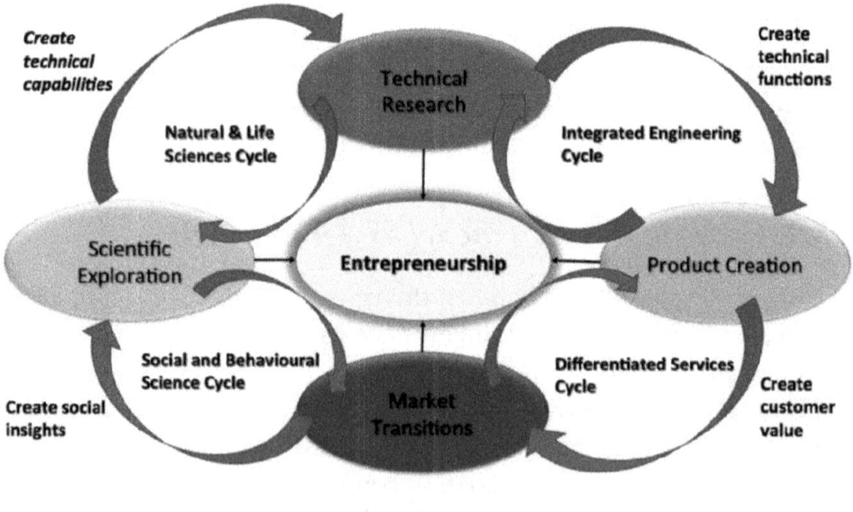

Figure 3.1. Cyclical model of technological entrepreneurship and innovation. *Source*: Trott *et al.* [2016, p. 20].

and Life Science Cycle creates technical capabilities. The Social and Behavioral Sciences Cycle creates social insights. Differentiated services cycle creates customer value. Integrated engineering cycle creates products [see Berkhout *et al.*, 2006; Trott *et al.*, 2016, p. 20]. Typically, patents are linked to natural and life sciences cycle and integrated engineering cycle. On the other hand, trademarks are linked to social and behavioral sciences cycle and differentiated services cycle.

> According to Berkhout, "Innovation projects must not be managed along the familiar linear pipeline but should be organized via cross-disciplinary networks along an innovation circle with ample internal feedback paths. Innovation may start anywhere on the circle and previous innovations will inspire new ones: innovations build on innovations. In such an organized chaos, causality is a meaningless concept and modern communication tools are indispensable. Experience shows that in innovation a shared mental framework is essential to allow synergy between the large number of highly diverse players. The Cyclic Innovation Model (CIM) is such a framework, being proposed to transform our current economy to a circular system: the circular economy".[6]

It is important to note that new technical capabilities and functions are in most cases created in innovation environments created and managed by science and technology parks, other innovation intermediaries and ecosystems. Also, new social insights and customer value are created. Entrepreneurs are not working in a sociocultural vacuum. They really need serendipity, platforms, and dynamic innovation ecosystems to further develop successful business ideas, inventions and innovations, whether they are technical innovations, business model innovations or social innovations. Of course, in the broad sense, value addition and brands are created for the consumers and end-users of products by entrepreneurs [Kunde, 2002]. In Table 3.3, we have presented a synthesis about the integration of the VUCA approach and the key cycles of technological entrepreneurship and innovation.

---

[6] URL: https://www.oecd.org/sti/inno/1874489.pdf.

Table 3.3. The integration of VUCA approach and key cycles of technological entrepreneurship and innovation.

	Serendipity management	**Innovation ecosystem**	**Platforms**
Social and behavioral science cycle	Increase diversity and improves the ability to benefit from unexpected	Creates gravity to pull new people and entities into the system	Offer a low entry barrier access to the system—even for global talent
Differentiated sciences cycle	Helps to identify also unexpected and contradictory differentiated services cycles	Makes service-dominant (S-D) logic available	Service as breeding environments for new and disruptive business models
Natural and life sciences cycles	Promotes transformations inside natural and life sciences cycle	Includes physical infrastructures helping to create 'sticky knowledge'	Host various scientific communities enabling multidisciplinary focus
Integrated engineering cycle	Enables the emergence of new combinations of technological competences	Provides rich catalytic element for attracting talent from diverse backgrounds	Competence platform enables cross-disciplinary idea elaboration engineering cycles stronger and more productive

Source: Kakko *et al.* [2016].

3.4 Evaluation Model for Vucability

The three solutions offered by this chapter—platform thinking, ecosystem development towards meta-ecosystems and serendipity management—will be fundamental in helping to succeed in the VUCA world. They have proved to be effective in the dynamic, global and non-linear innovation environment of Aalto University campus in Espoo, Finland, and the Urban Mill case has been presented in more detail in two WTA/UNESCO Training Workshop papers [Kakko and Mikkelä, 2015; Kakko, 2017]. This encourages us to present an evaluation model, which will help all innovation intermediary entities—and especially STPs—to understand and improve their preparedness for future challenges. There are three critical dimensions to be considered when adapting into the circumstances of the Postnormal VUCA era: (1) the density of serendipity thinking, (2) the maturity of innovation ecosystems and (3) the depth of platform thinking. We can visualize scenarios for innovation intermediaries following (see Figure 3.2).[7]

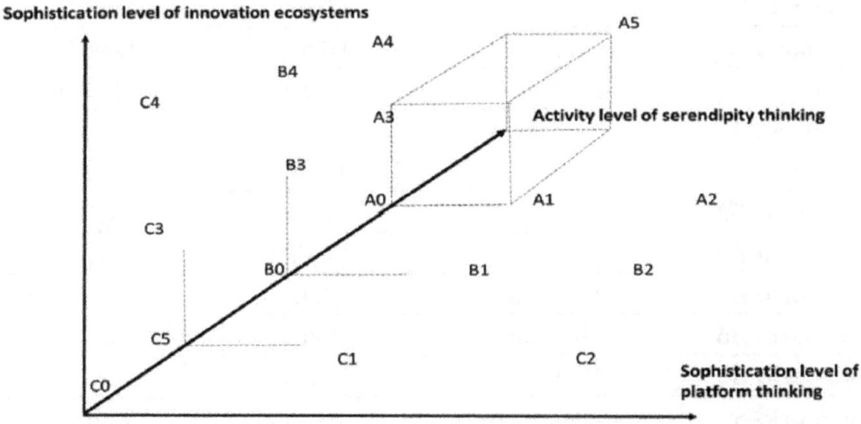

Figure 3.2. Vucability improvement model for innovation intermediaries like ecosystems and STPs.
Source: Kakko *et al.*, 2016.

[7] The figure looks as it does in the original paper, since we have changed the notions as indicated above.

Table 3.4 presents the characteristics of potential STP management scenarios by the vucability criteria. We can note that there are various development paths toward excellence in this evolutionary vucability process. This kind of scenario space map of alternative futures is one way to understand complex adaptive systems and potential futures maps and path developments [see Malaska and Virtanen, 2005]. This kind of scenario approach is often used in morphological system analyses [Ritchey, 2011].

Table 3.4. Potential scenarios for vucability improvement.

	Activity level of serendipity thinking (including density evaluation)	Sophistication level of innovation ecosystems	Sophistication level of platform thinking
Scenario C0	Low	Low	Low
Scenario C1	Low	Low	Medium
Scenario C2	Low	Low	High
Scenario C3	Low	Medium	Low
Scenario C4	Low	High	Low
Scenario C5	Low	Medium	Medium
Scenario B0	Medium	Low	Low
Scenario B1	Medium	Medium	Medium
Scenario B2	Medium	Medium	High
Scenario B3	Medium	Medium	Low
Scenario B4	Medium	High	Low
Scenario B5	Medium	High	High
Scenario A0	High	Low	Low
Scenario A1	High	Medium	Medium
Scenario A2	High	Medium	High
Scenario A3	High	Medium	Low
Scenario A4	High	High	Low
Scenario A5	High	High	High

The highest evolutionary level of the vucability is Scenario A5, where the density and activity level of serendipity thinking, the sophistication level of ecosystem development, also known as the maturity of innovation ecosystem, and the sophistication level of platform thinking, also known as the depth of platform thinking, reach the highest possible levels. Naturally, the lowest level of the viability of Scenario C0, where the density level of serendipity thinking, the maturity of innovation ecosystem and depth of platform thinking reach only lowest evolutionary levels, show here a preliminary framework on how the VUCA index can be applied for the improvement of STP activities. Further research based on new case studies is ongoing, and practical results and insights to define the correct and viable criteria and ratings are created all the time.

3.4.1 *Building an evaluation and mapping system for vucability*

This idea about evaluation and mapping system for vucability was first introduced in an article and presented in a keynote lecture at Tangerang-Selatan Global Innovation Forum 2016 [Kakko *et al.*, 2016]. At that time, the world was very different and the need for such a system was not apparent. Our research and business interests were directed by other domains, the focus was mainly in how improving the platform thinking and research will transform the global business climate. The ignition to take vucability again into our agenda was working with several interesting customer cases where new elements for our vucability framework were discovered (see Cases 1, 2 and 3).

We have worked to study and develop this concept further alongside our normal consultancy business; however, it has become mainstream only recently. While introducing this approach to some of our collaboration partners in Europe, we have been running analysis and finding new conclusions. So, in the last three years' time, we have run some experiments, but unfortunately, we cannot identify the cases in more detail here because of our mutual agreements and the highly confidential and sensitive information involved. However, we will briefly describe them case by case.

Case 1. Meta Ecosystem—Putting Together a Consortium

In this case, we worked with an assignment to put together a consortium of ecosystems with the ultimate goal of activating the collaboration with the participating ecosystems. We interviewed applicants and selected five domain-specific ecosystems and two other ecosystems as multipliers from seven EU countries. Together, we created a strategic plan and a roadmap for implementation. In this process, our understanding of the problems of domain-specific and siloed ecosystems in Europe crystallized. As a result, we introduced the idea and roadmap for a first European level meta-ecosystem. The process is now ongoing following the established roadmap.

Case 2. Meta Ecosystem—Building a Node

In this case, we collaborated with the government of a Small Island State and discovered the deep concerns of their export-driven and not diverse-enough economy. In developing countries, the globalization might hit hard and cause brain drain. We elaborated on the issue together with the government and their science park management team, and ended up with a solution, which is based on building a node there that is tightly integrated to a meta-ecosystem. Establishing a physical space, a node, supported by a competence platform, we can attract foreign, mainly European freelancers, startups, and SMEs to come and stay for longer periods of time in the tropical island and find the benefits of a changing environment.

Case 3. Competence Platform—Improving Functionalities

We have worked closely with several competence platform providers. This development work and the pilots we have run has helped us to understand the elementary functionalities such a platform needs when applied in an innovation intermediary environment. The platform classification where the role of competence platform within a platform universe is illustrated in more detail in our paper titled "Platform

Thinking within the Third Generation Science Park Concept" [Kakko and Mikkelä, 2015].

The main functionalities of a competence platform are (1) user profiles, both skills and wills—from members, (2) free idea elaboration with reference to those skills, (3) team formation based on notifications and references and (4) various features for further idea elaboration. So, competence platforms are not only communication platforms like Slack or Teams; instead, they go a step further in self-orchestrating the work of an ecosystem operator.

Case 4. Social Media—Flow of Information Filtered by a Diversified Community

In this case, we used Facebook in order to put together a Facebook group, and hence a community also interested in surviving and prospering in the bear market conditions of the global stock exchanges. The community grew immediately with high interest; in the first week, more than 1,000 members joined the group. This case has proven that gravity creates gravity and pull. The dynamics in the group has indicated that in the VUCA investment environment, when information flow is massive and interpretation of the data is quick and diverse, immediate peer support and the effort of many experts to filter the most relevant information and keep the community updated is highly appreciated. With this case, we are continuously drawing new insights about an ability of a virtual community to adapt in VUCA conditions much better than each and every one of them working independently.

Based on this work with customer cases, we have outlined the most important factors for all these three dimensions of vucability. In Tables 3.5 to 3.7, we present some examples from our database.

The categorization of qualitative variables presented in this chapter may help STP managers and ecosystem operators to understand their vucability level in a world hit by the coronavirus pandemic. We have mobilized a research and development team for elaborating the evaluation criteria further with more use cases. The ultimate goal is getting it conceptualized, and we are looking for collaboration partners both for the technical development of analysis and consultancy and beta-level customers like STPs and ecosystem operators. The

Table 3.5. Density of serendipity thinking.

Evaluation criteria	Low	Medium	High
Level of cyclic innovation model (CIM) skills and capacities	Closed innovation management	Open innovation management	CIM excellence
Overall diversity of the ecosystem	Low	Medium	High
Local versus global approach	Local	National	Global
The overall ratio of free agents (students, freelancers, startups and researchers) in ecosystem	Low Less than 1:10	Medium Less than 1:3	High More than 1:3
The ratio of gatekeepers, intermediary managers and facilitators in ecosystem	High More than 1:20	Medium	Low Less than 1:100
The use of competence platform in team formation (% of team founded)	None 0%	Passive use 1–25%	Active use 25–100%
Overall respect for serendipity on individual, community and ecosystem levels (evaluated with a list of selected questions)	Low Score below 5	Medium Score 5–10	High Score 10+

Table 3.6. Transformation towards Meta Ecosystem.

Evaluation criteria	Low	Medium	High
The variety of orchestration methodologies	1–2	3–4	4+
The diversity of the communities within the ecosystem	Low	Medium	High
The number of interdependent stakeholders	High	Medium	None
The ratio of gatekeepers, intermediary managers and facilitators in ecosystem	High More than 1:20	Medium	Low Less than 1:100
The collaboration activity level with other ecosystems	Low Less than 5%	Medium 5–15%	High 15+ %
The use of competence platform in team formation (% of team founded)	None 0%	Passive use 1–25%	Active use 25–100%

Table 3.7. The depth of platform thinking.

Evaluation criteria	Low	Medium	High
The utilization of various platforms	1–2	3–4	4+
The gravity of platform, i.e., the critical mass of active users and the growth rate	Low	Medium	High
The number of participants outside the proximity	None	Medium	Majority
The level of self-organized and peer-to-peer curated initiatives	Low Less than 1:10	Medium Less than 1:3	High More than 1:3
The users outside ecosystem	Low Less than 5%	Medium 5–15%	High 15 + %
The use of competence platform in team formation (% of teams founded)	None 0%	Passive use 1–25%	Active use 25–100%

Table 3.8. Planning in Normal Era to Postnormal Era.

Normal era	**Postnormal era**
Productivity, Profitability	Vucability
Push → Top-down	Pull → Gravity
Command and control	Connectivity and collaboration
MBA driven management	Emergent leadership
Linear planning	Nonlinear planning
Strategic planning	Agility, ability to implement
Pipeline thinking	Platform thinking
Project management	Serendipity management
Venture capital	Social capital
Networking	Community building
Corporations and startups	Diverse forms of entrepreneurship
Domain specific ecosystems	Meta ecosystems
Local and national problems	Wicked global problems, SDGs
Incremental economy	Sharing economy
Predictability	Gaining from unexpected
Hierarchy	Wirearchy
Stoke of information	Flow of information
Decision making	Sense making
Shareholder value	Stakeholder value
Ownership	Access
Human resources on organization's payroll, office work	Tapping in the global talent pools, flash teams, remote working
Motivation	Passion
Wealth	Health
Work-life balance (measured 24/7)	Work-life balance (measured during the lifetime)
Fragile	Antifragile
Dualism	Rhizome

actual measures to improve the maturity of this approach, and therefore the capability to succeed in the Postnormal era will be conveniently determined, when further studies on the evaluation and selection of valid and useful criteria is conducted.

3.5 Final Conclusions

The main conclusion in this chapter is that the world after the coronavirus crisis will be fundamentally different from the Normal era we have lived in for so long. The transformation seems to become quick and violent, and serious re-judgments of processes, social and economic behaviors and values are needed. We would like to conclude the revolutionary move from the Normal era to the Postnormal era in Table 3.8, which is an updated version of the table first published in my book [Kakko, 2014]. The focus in Table 3.8 is in the changes of organizational and individual behavior, and technology changes are not included. We believe that understanding this fundamental shift will help businesses and societies to better prepare for this huge transformation, when we have to enter the Postnormal era with all the wicked problems following.

The future ain't what it used to be.

–Yogi Berra

References

Berkhout, A.J., Hartmann, D., van der Duin, P., and Ott, R. (2006). Innovating the innovation process. *International Journal of Technology Management, 34*(3–4), 390–404.

Boyd, S. (2013). *How the Postnormal Era will Change Everything*. URL: http://www.stoweboyd.com/post/49946797759/how-the-postnormal-era-will-change-everything. Accessed: August 9, 2016.

Carayannis and Campbell. (2009). 'Mode 3' and 'Quadruple Helix': Toward a 21st century fractal innovation ecosystem, *International Journal of Technology Management, 46*(3/4).

Camarinha-Matos, Afsarmanesh, and Ollus (eds.). (2008). *Methods and Tools for Collaborative Networked Organizations*. New York: Springer.

Choudary, S.P., van Alstyne, M., and Parker, G. (2013). The rise of the platform: A seismic shift in business models, *The Definitive Starting Point to Understanding Why Platforms Are Eating the World*. URL: https://www.dropbox.com/s/pwenw882lgx07oz/The-Rise-of-the-Platform-A-seismic-shift-in-business-models.pdf?dl=0. Accessed: August 7, 2016.

Etzkowitz, H. (1994). Academic-industry relations: A sociological paradigm for economic development, in: Leydesdorff, L., and van den Besselaar, P. (eds.), *Evolutionary Economics and Chaos Theory: New Directions in Technology Studies*, Pinter, London, etc., pp. 139–151.

Etzkowitz, H., and Leydesdorff, L. (1995). The Triple Helix—University–Industry–Government relations: A laboratory for knowledge-based economic development, *EASST Review*, *14*(1), 14–19.

Jackson, D.J. (2011). What is an innovation ecosystem?, *National Science Foundation*, Arlington, VA. URL: http://erc-assoc.org/sites/default/files/topics/policy_studies/DJackson_Innovation%20Ecosystem_03-15-11.pdf.

Kakko, I. (2014). *Oasis Way and the Post-Normal Era–How Understanding Serendipity Will Lead You to Success*, St. Petersburg BHV, St. Petersburg, Russia.

Kakko, I. (2017). How to improve the collaboration of various stakeholders in VUCA environments by platform thinking and ecosystem development, WTA/UNESCO International Training Workshop, September 2017, Daejeon, South Korea.

Kakko, I., and Inkinen, S. (2009). Homo creativus: Creativity and serendipity management in third generation science and technology parks, *Science and Public Policy*, *36*(7), 537–548.

Kakko, I., and Mikkelä, K. (2015). *Platform Thinking within the Third Generation Science Park Concept*, WTA/UNESCO International Training Workshop, September 2015, Daejeon, South Korea.

Kunde, J. (2002). *Unique Now or Never. The Brand is the Company Driver in the New Value Economy*, London: Prentice Hall.

Kakko, I., Kaivo-oja, J., and Mikkelä, K. (2016). How to support and develop the innovation-oriented entrepreneurship in Turbulent VUCA conditions? *Keynote at Tangerang–Selatan Global Innovation Forum*, organized by UNESCO and WTA, September 2016, Indonesia.

Kakko, I. (2017). How to improve the collaboration of various stakeholders in VUCA environments by platform thinking and ecosystem development, WTA/UNESCO International Training Workshop, September 2017, Daejeon, South Korea.

Kakko, I. (2010). Oasis Way and the Postnormal Era—How understanding serendipity will lead you to success, *BHV-St. Petersburg*, ISBN 978-5-9775-3347-8, p. 11.

Lowe, C.U. (1982). The Triple Helix-NIH, industry, and the academic world, *The Yale Journal of Biology and Medicine*, *55*, 239–246.

Malaska, P., and Virtanen, I. (2005). Orienteering in the futures universe: A map-analogy-based set-theoretic approach to the theory of futuribles, in: Laitinen, E.K., and Laitinen, T. (eds.), *Contributions to Accounting, Finance and Management Science, Essays in Honour of Professor Timo Salmi*, Acta Wasaensia No. 143, pp. 261–284.

Merton, R.K., and Barber, E. (2004). *The Travels and Adventures of Serendipity—A Study in Sociological Semantics and the Sociology of Science*. Princeton, New Jersey, USA: Princeton University Press.

Oh, D.S. Phillips, F., Park, S., and Lee, E.H. (2016). Innovation Ecosystems: A critical examination, *Technovation*, *54*, 1–6.

Raita, E. (2020). Meta-ecosystem as a tool to generate cross-disciplinary innovations with global impact. URL: http://www.karostech.fi/meta-ecosystem-as-a-tool-to-generate-cross-disciplinary-innovations-with-global-impact/.

Ravetz, J.R. (1993). Science for the post-normal age, *Futures*, *25*, 735–755.

Ravetz, J.R., and Funtowicz, S.O. (1999). Post-normal, science: An insight nowmaturing, *Futures*, *31*(7), 641–646.

Rabelo, R.J., and Bernus, P. (2015). A holistic model of building innovation ecosystems, *IFAC-PapersOnLine*, *48*(3), 2,250–2,257.

Ritchey, T. (2011). Modelling Alternative Futures with General Morphological Analysis, *World Future Review*. Spring.

Sardar, Z. (2015). Postnormal times revisited, *Futures*, *67*(2015), 26–39.

Sardar, Z., and Sweeney, J.A. (2016). The three tomorrows of postnormal times, *Futures*, *75*, 1–13.

Sardar Z. (2010). Welcome to postnormal times, *Futures*, *42*(1), 435–444.

Sábato, J., and Mackenzi, M. (1982). *La Producción de Technologia. Autónoma o Transnacional*. Mexico: Nueva Imagen.

Teece, D.J. (1986). Profiting from technological innovation: Implications for integration, collaboration, licensing and public policy, *Research Policy*, *15*(6), 285–305.

Teece, D.J. (2006). Reflections on Profiting from Innovation, *Research Policy*, *35*(8), 1,131–1,146.

Trott, P., Hartmann, D., van der Duinn, P., Scholten, V., and Ortt, R. (2016). *Managing Technology Entrepreneurship and Innovation*, p. 20. London and New York: Routledge.

URL: http://www.thunderbird.edu/article/living-and-leading-vuca-world. Accessed: 8.8.2016.
URL: http://www.aj-berkhout.com/Cyclic%20Innovation%20Model/.

Acknowledgements

This chapter is an updated and upgraded conclusive version of the following papers and presentations:

Kakko, I., and Mikkelä, K. (2015) Platform Thinking within the Third Generation Science Park Concept, WTA/UNESCO International Training Workshop, September 2015, Daejeon, South Korea.
Kakko, I., Kaivo-oja, J., and Mikkelä, K. (2016). How to support and develop the innovation-oriented entrepreneurship in Turbulent VUCA conditions? *Keynote at Tangerang–Selatan Global Innovation Forum*, organised by UNESCO and WTA, September 2016, Indonesia.
Kakko, I. (2017). How to improve the collaboration of various stakeholders in VUCA environments by platform thinking and ecosystem development, *WTA/UNESCO International Training Workshop*, September 2017, Daejeon, South Korea.
Kakko, I. (2018). *The Role of Research and Development, Platform Thinking and Serendipity in Orchestration of an Innovation Ecosystem*, UNESCO/KOICA training for STP Managers, April 2018, Jakarta, Indonesia.

© 2023 World Scientific Publishing Company
https://doi.org/10.1142/9789811257186_0004

Chapter 4

Understanding GARLICs: An Innopreneurship Platform for Smart Cities

Arcot Desai Narasimhalu

Department of Information Systems, School of Computing and Information System, Singapore Management University, Simha Innovations Pte. Ltd., Singapore
Email: desai@smu.edu.sg

4.1 Introduction

This chapter examines the elements required to create an innovation-driven entrepreneurship platform for smart cities. The examples are drawn from the US, Israel, and Singapore, and therefore its coverage is neither comprehensive nor complete. The US is chosen because it has continuously launched several new industry-defining innovations whether technology-driven or otherwise. Israel is chosen since it is generally branded as "The Startup Nation", and is a prime example of a small and savvy country that has successfully scaled their startups into the American markets. Singapore is chosen because it places among the top 10 in a number of smart city rankings. There are also a few notable examples drawn from some other countries as well. In this chapter, the term "innopreneurship" refers to innovation-based entrepreneurship. Innopreneurs are individuals who successfully identify, develop and commercialize innovative products, services or processes. Innopreneurs need to manage market, technology and business risks, whereas traditional business leaders focus largely on managing their business risk.

The acronym 'GARLICs' denotes the seven actors of an innopreneurship platform—Government, Academia, Research Labs, Large enterprises, Investors, Citizens, and small- and medium-sized enterprises. Section 4.2 of this chapter introduces the concept of smart cities and innovation platforms. The dynamics of an innopreneurship platform are discussed in Section 4.3. Section 4.4 discusses the GARLICs platform in the context of smart cities and offers a means of calculating a score for a sustainable smart city. The critical actors and programs that can help build a sustainable smart city are discussed in Section 4.5. Section 4.6 summarizes the discussions in the rest of the chapter.

4.2 Smart Cities and Innovation Platforms—A Recap

In this section, we introduce smart cities and innovation platforms both as an introduction to first timers and as a refresher to those who are already aware of the concepts.

4.2.1 *Smart city*

A smart city is a designation given to a city that incorporates information and communication technologies (ICT) to enhance the quality and performance of urban services such as energy, transportation and utilities in order to reduce resource consumption, wastage and overall costs. The overarching aim of a smart city is to enhance the quality of living for its citizens by leveraging smart technologies.

4.2.1.1 *A sample smart city project*

The Brussels smart city project is an example of a smart city [Brussels]. The wheel shown in Figure 4.1 is a representation of Brussels' smart city projects that was inspired by the definition of a smart city formulated by the researcher Anthony Townsend of New York University—"A smart city is a place where ICTs are combined with infrastructure, architecture, daily objects and even our body, in order to tackle social, economic and ecological problems". Such a definition implies that the objective of any smart city ought to be driven by the Triple Bottom Line (TBL) framework—Financial, Social and Environmental, as shown in Figure 4.2.

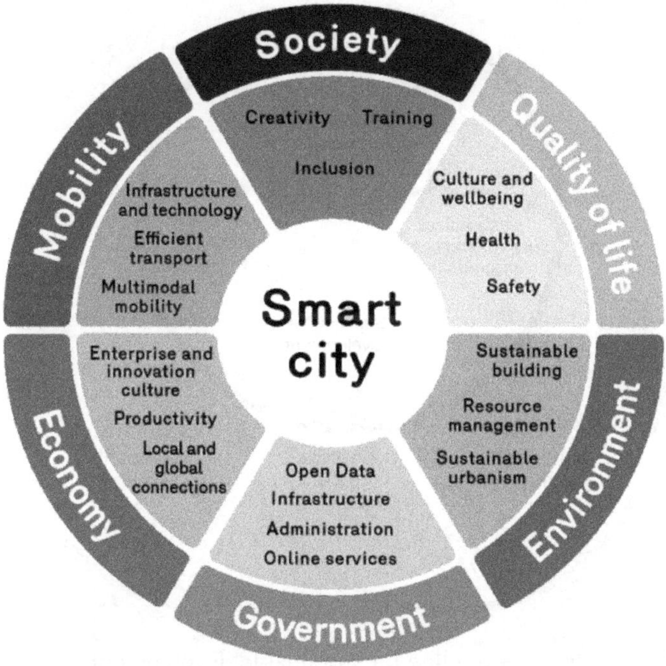

Figure 4.1. Brussel's smart city wheel.
Source: Brussels Smart City.

The Brussels smart city projects define the following as their objectives:

(1) Sustainable development in response to ecological issues.
(2) Human development in response to social issues.
(3) Economic development in response to the issues of prosperity.

These three objectives are tightly aligned with the Triple Bottom Line framework introduced earlier. Brussels' smart city projects uses the following six action items to achieve their objectives:

(1) *Collaborate*: Public services should operate in an open environment in a smart city. They must share their data and projects to ensure that citizen-focused services are more efficient and effective while achieving economies of scale.

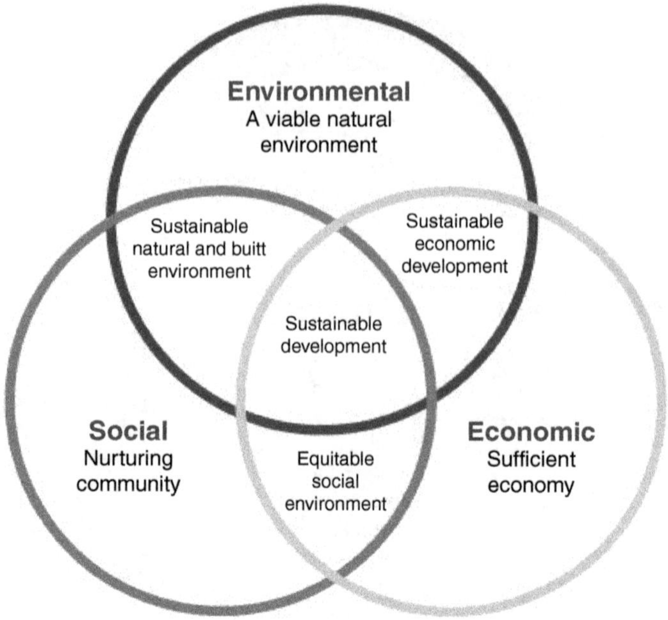

Figure 4.2. Triple bottom line for sustainable development.
Source: The Triple Bottom Line, by Hall, Jenkins, Kearsley (1997) in *Global Tourism*.

(2) *Save*: Smart cities should aim to reduce the use of energy, water, raw materials, food, financial resources and polluting emissions in order to achieve a sustainable future.
(3) *Innovate*: Smart cities should establish policies and programs to encourage its citizens, businesses and public services to come up with new ways of organizing, sharing, communicating and producing products and services for a sustainable future.
(4) *Integrate*: Smart cities' policies and programs ought to reduce social divisions in areas such as education, gender, health and safety.
(5) *Participate*: Smart cities must involve its citizens and businesses in planning projects for achieving a sustainable future.
(6) *Simplify*: Smart cities remove and revise regulations or customs that are either obsolete or counterproductive and often out-of-phase with the simplicity of web-based services.

Brussels' smart city projects use the following three-layer technology architecture to achieve their objectives:

(1) The lowest layer, continually growing *digital data* is a smart city's raw material.
(2) The middle, or *infrastructure*, layer interconnects all the necessary equipment to offer access to available data.
(3) The top, or *smart*, layer creates information from the data to be accessed and used by government, citizenry and businesses.

4.2.1.2 *Components of a smart city*

Boyd Cohen's Smart City Wheel, as shown in Figure 4.3, represents six components of a smart city that has been widely embraced by several smart city projects [Cohen, 2014].

The six components of a smart city as defined by Cohen's Smart City Wheel are:

(1) *Smart economy*: Smart cities should attract and use talent to build an innovation-based economy. Smart cities ought to understand and relate to the need to reengineer their economies on a regular basis in order to avoid stagnant economies.
(2) *Smart governance*: Smart cities should offer digital technology and digital data-powered public services.
(3) *Smart environment*: Smart cities should offer environmentally friendly living spaces that reduce the consumption of energy and natural resources and minimize pollution.
(4) *Smart mobility*: Smart cities should offer the benefits of low congestion and pollution by promoting collective and sustainable travel.
(5) *Smart population*: Smart cities should train their citizens to seamlessly transition from one economy to the next. Citizens should also have access to information that will help them design and organize a sustainable lifestyle.
(6) *Smart living environment*: Smart cities should offer the highest levels of healthy and safe living.

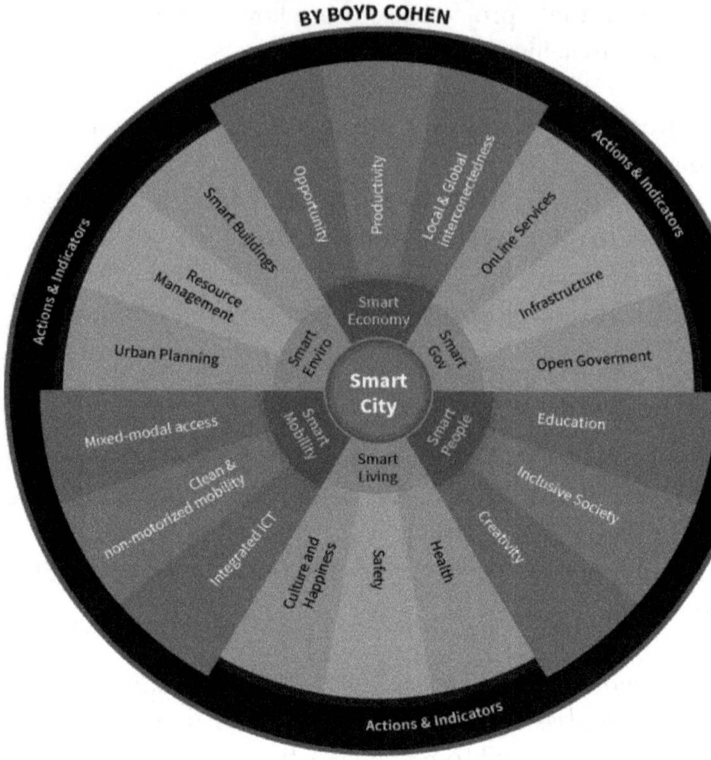

Figure 4.3. Boyd Cohen's smart city wheel.
Source: Cohen [2014].

4.2.1.3 *Smart city rankings*

There are several smart city rankings, each using a different set of metrics. It is however important to note that some cities consistently appear in the top 10 of all notable rankings.

In 2017, Juniper Research [Burggarf, 2018] used the Intel-sponsored Global Smart City Performance Index to rank top 20 cities that successfully integrated Internet of Things (IoT) technologies and connected services across mobility, healthcare, public safety and productivity to give back more time to the citizens of a city. These were Singapore, London, New York, San Francisco, Chicago, Seoul, Berlin, Tokyo, Barcelona, Melbourne, Dubai, Portland, Nice, San Diego, Rio de

Janeiro, Mexico City, Wuxi, Yinchuan, Bhubaneswar and Hangzhou in that order.

In 2017, EasyPark [EasyPark] group also did a survey of 500 cities, 200 of them considered to be smart, derived the metrics that would define a smart city and then ranked all the cities. They listed Copenhagen, Singapore, Stockholm, Zurich, Boston, Tokyo, San Francisco, Amsterdam, Geneva and Melbourne as the top 10 smart cities based on 19 factors.

In July 2018, Forbes [IESE Business School, 2018] carried an article from an IESE Business School contributor that ranked New York, London, Paris, Tokyo, Reykjavik, Singapore, Seoul, Amsterdam, and Hong Kong as the 10 smartest cities of the world. The study used the IESE Cities in Motion Index.

However, an important aspect of a smart city is its ability to regularly reengineer its aging economy. This aspect has not been included explicitly in smart city rankings. A metric that includes a smart city's ability to reengineer its economy is important and is discussed later.

4.2.2 *Innovation platforms*

Innovation platforms caught the attention of development agencies at the end of the 1980s, when the private sector started using such platforms to gather information and improve networking among its key stakeholders. Innovation platforms are increasingly common in research and development initiatives.

4.2.2.1 *Definition of an innovation platform*

The UK government identifies an innovation platform as a space for learning and change [GOV.UK, 2013]. It defines an innovation platform to be a group of individuals (who often represent organizations) with different backgrounds and interests—farmers, agricultural input suppliers, traders, food processors, researchers, government officials, etc. Its members come together to develop a common vision and find ways to achieve their goals. They may design and implement activities as a group or coordinate activities by individual members. Individual members can also innovate alone, spurred on by the coordinated group activities.

4.2.2.2 *Extending the participants in an innovation platform*

Many smart city projects such as Brussels' smart city projects define a platform to be made up of four main actors—(1) government (i.e., public services), (2) citizens, (3) businesses and (4) academia (i.e., universities). While this is a good starting point, there is a need to examine some of the actors in detail.

Businesses should be further subdivided into large businesses, small and medium businesses and investors. Each category of businesses plays a unique role in a smart city and they will require targeted policies from governments. Non-profit organizations such as research institutes should also be considered as a separate category of business, given their unique nature and role. The rationale for subdividing businesses as an actor is to provide for policies and programs that are relevant for different subcategories of business.

Universities should be generalized to academic institutions to additionally include secondary educational institutions (grades 7 to 12), if not even primary schools (grades 1 to 6). The reason for expanding educational institutions beyond universities is to instill the spirit, knowledge and mindset for entrepreneurship and innovation from the earliest possible ages.

With these changes, the stakeholders involved in an innovation-enabled entrepreneurship platform will be:

- Government
- Academic institutions—primary, secondary and tertiary levels
- Research institutions that focus on translational efforts
- Large enterprises that include both local and multinational corporations
- Investors including angel investors and crowd equity funding communities
- Citizens
- Small- and medium-sized enterprises that include startups.

These conveniently abbreviate into GARLICs. A lower-case letter 'S' is used to represent the small- and medium-sized enterprises. GARLICs is an extension of the term 'CUGAR' (Companies Universities

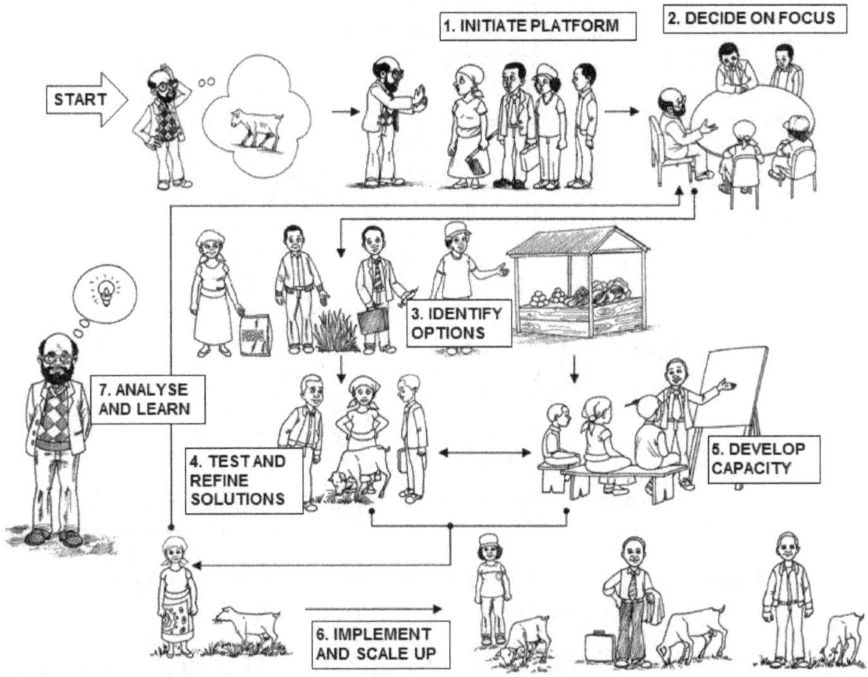

Figure 4.4. The seven-stage process used by innovation platforms.
Source: GOV.UK, 2013.

Government Angel investors and venture capitalists, and Research labs) that was developed as an innovation platform for Science and Technology Parks [Narasimhalu, 2012].

Innovation platforms can be set up to tackle challenges and opportunities at various levels—in a village or community, in a district or nationwide, or throughout a value chain or economic sector. A typical seven-stage process deployed by innovation platforms is shown in Figure 4.4.

4.3 The Dynamics of an Innopreneurship Platform

An innopreneurship platform is an innovation platform that is entrepreneurial in nature or is run by entrepreneurs. Intensive and comprehensive participation and contributions to and from all actors in an innopreneurship platform is a requirement for it to be the most

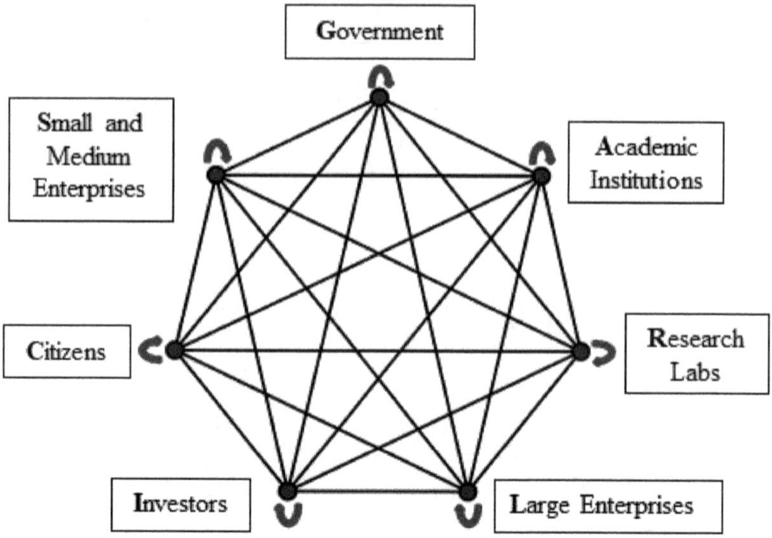

Figure 4.5. Innovation platform actors interaction graph.

effective. Such contributions can take the form of policies, programs, recommendations and guidance.

The interaction between each of the seven actors and the rest in an innopreneurship platform is represented as a total graph as shown in Figure 4.5. Each link in the graph represents two actors and their contributions to each other. The reflexive link denotes the renewal and reinventing of all seven actors. The contribution of each of the actors is captured in Tables 4.1 to 4.7. In these tables, existing policies and programs are represented in normal typeface with attribution to the country that has implemented them while policies and programs that should be introduced are represented in italics. Some of the programs in italics may exist and are either invisible or not known widely. Programs and policies that are probably practiced but cannot be easily attributed are represented in normal typeface.

It is important to design a metric to determine the quality of an innopreneurship platform. This is achieved by assigning scores or weights to links connecting the different actors in GARLICs. A score of +1 represents maximum positive contributions, a score of 0 represents neutral or no contribution, and a score of –1 represents maximum

Table 4.1. Sample programs and policies that a government can offer to other innopreneurship platform actors.

Contributor	Beneficiary	Components or programs
Government	Government	FAST: Federal and State Technology Partnership (USA) BIRD: Binational Industrial R&D program (USA, Israel) SIIRD: Singapore Israel Industrial R&D program (Israel, Singapore) Repurposing older resources for emerging industry–Euratechnologies (France)
	Academic institutions	STTR: Small Business Technology Transfer (USA) Sophia Antipolis (France) UIF: University Innovation Funds (Singapore) YES: Young Entrepreneurs' Scheme (2008 Singapore) ACE: Action Community for Entrepreneurship Scheme (2013 Singapore)
	Research Labs	STTR: Small Business Technology Transfer (USA) Sophia Antipolis (France)
	Large Enterprises	Generic Technology Program (Israel) Sophia Antipolis (France) PIC: Productivity and Innovation Credit Scheme (Singapore) EDB Investments (Singapore) Enterprise Development Grant (Singapore) Productivity Solutions Grant (Singapore)

(*Continued*)

Table 4.1. (*Continued*)

Contributor	Beneficiary	Components or programs
	Investors	YOZMA (Israel)—created Venture Capital Industry [YOZMA] TIS: Technology Incubator Scheme (Singapore) ESVF: Early Stage Venture Fund (Singapore) SEEDS Capital (Singapore) AITD: Angel Investor Tax Deduction (Singapore) FMI: Fund and Fund Management Incentives (Singapore)
	Citizens	Skills Future (Singapore) *Minds Future*
	Small- and Medium-sized Enterprises	SBIR: Small Business Innovation Research (USA) [SBIR] STTR: Small Business Technology Transfer (USA) [SBIR] TIBBETS Award: Top SBIR performers (USA) [SBIR] BIRD: Binational Industrial R&D program (USA, Israel) Generic Technology Program (Israel) [BIRD] Industrial and Technology Incubators (Israel) Sophia Antipolis (France) [Sophia] PIC: Productivity and Innovation Credit Scheme (Singapore) Startup SG Founder Incubation Grant (Singapore) [Enterprise SG] Startup SG Tech (Singapore) Enterprise Development Grant (Singapore)

Table 4.1. (*Continued*)

Contributor	Beneficiary	Components or programs
		Productivity Solutions Grant (Singapore)
		SME Micro Loan (Singapore)
		SME Working Capital Loan (Singapore)
		SME Venture Loan (Singapore)
		SME Equipment and Factory Loan (Singapore)
		IFS: Internationalization Financing Scheme (Singapore)
		LIS: Loan Insurance Scheme (Singapore)
		PRIS: Political Risk Insurance Scheme (Singapore)
		SG Innovate: Deep technology business building (Singapore)
		TCIS: Trade Credit Insurance Scheme (Singapore)

negative contribution. Using these scores, the maximum any actor can earn is 7 and the minimum score that the actor can earn is –7. Table 4.8 gives an approach to scoring each of the links.

The maximum raw score for the quality of an innopreneurship platform will be 49 and the minimum raw score will be –49. The normalized innopreneurship platform's score (NIPS) can be calculated by the formula (raw score + 49 + 2)/10. 2 is added to produce a rounded off sore. The NIPS of a smart city can range between 0 and 10.

4.4 Innovation Platforms Applied to a Smart City

A close examination of smart cities will reveal that they have one irrefutable trait—their ability to reinvent or reengineer their economies on

Table 4.2. Academic institutions' sample contributions to other innopreneurship platform actors.

Contributor	Beneficiary	Components or programs
Academic Institutions	Government	*Help train innovation-minded civil service* *Propose developing innovative and productive citizen services*
	Academic institutions	Karolinska Innovation's START and DRIVE (Sweden) [KI] Develop Innovation and Entrepreneurship related academic programs (SMU/NTU, Singapore) Establish entrepreneurial culture (SMU/NUS, Singapore) YIC: Youth Innovation Competition for secondary schools (SMU, Singapore) AYEP: Asian Youth Entrepreneurship Program for cross-country entrepreneurship programs (SMU, Singapore)
	Research Labs	HADASSIT Hebrew University's medical commercialization company (Israel) WARF: Wisconsin Alumni Research Foundation (USA) [WARF]
	Large Enterprises	Industry leaders as mentors and instructors (IIT Madras India) *Innovation and Innopreneurship-oriented Executive Development Programs* (*Universities*?) Innovation subcontracts (SMU-X Bank, Singapore)
	Investors	Karolinska Development (Sweden) [KD]
	Citizens	Innopreneurship focused continuing education programs (ISS/NUS, Singapore)
	Small- and Medium-sized Enterprises	*Innopreneurship Partnership programs* Innovation catalyst programs (SMU, Singapore) *Adjacent innovations catalyst program*

Table 4.3. Research labs' sample contributions to other innopreneurship platform actors.

Contributor	Beneficiary	Components or programs
Research Labs	Government	*Guidance on science and technology trends*
	Academic institutions	*Guidance on market opportunities* *Translational projects around promising scientific and technological inventions*
	Research Labs	*Collaborative projects for solving big problems*
	Large Enterprises	*Offer innovative market ready product/ service prototypes*
	Investors	*Offer investment opportunities in first of a kind innovation*
	Citizens	*Meet the citizen session to inspire latent and budding innovators and entrepreneurs*
	Small- and Medium-sized Enterprises	*Develop next-generation innovation as an innovation outsourcing partner*

a regular basis. For example, Singapore started with a manufacturing economy and then moved into IT and finance industries to rejuvenate its economy. She is at present heading full steam transitioning into a knowledge-based and innovation-enabled economy.

A stellar example of a smart city/region is the Silicon Valley. They have used startups to fuel the stability and growth of their region by regularly creating new industries. Silicon Valley started with electronics, moved into Internet including e-commerce, search and other subindustries, transitioned into Biopharmaceuticals and is currently creating an economy fueled by Artificial Intelligence (AI), Big Data and Blockchain technologies. The main driving force that enabled such transitions was not just its regional governments. It was indeed the diverse entrepreneurial talent and enlightened investment community that helped create several new industries.

Table 4.4. Large enterprises' selected contributions to other innopreneurship platform actors.

Contributor	Beneficiary	Components or programs
Large Enterprises	Government	*Guidance on market and industry dynamics* *Guidance on skill sets required for future markets and industries*
	Academic institutions	Establish Joint R&D Labs (Intel's Lablets) *Create Translational R&D Labs* *Guidance on skill sets required for future workforce*
	Research Labs	*Inputs on Market demands*
	Large Enterprises	Initiate and Foster Innovation Culture (3M, P&G, Apple, Google) Establish Open Innovation Labs (P&G) Establish joint ventures (Sony, Ericsson)
	Investors	Establish Venture Funds (Nokia Ventures, Intel Capital) Establish partnerships with Venture Capital [CISCO]
	Citizens	*Sponsor innovation competitions* (*Abbott Labs*) *Hire individuals with innovation mindset*
	Small- and Medium-sized Enterprises	*Provide market and product development expertise as a means to later acquire or establish joint venture the SME*

The key, if not the only, indicator for the sustainability and growth of a smart city is its ability to stabilize or grow its economy by embracing, fostering and helping grow new industries. None of the smart city definitions or rankings tend to capture this important indicator. A smart city that does not proactively use innopreneurship platforms to create disruptions to its economy will end up producing incremental innovations that will either stagnate or shrink its economy.

Table 4.5. Values that investors can offer to other innopreneurship platform actors.

Contributor	Beneficiary	Components or programs
Investors	Government	Use pension and other funds to offer better returns Co-sponsor innovation and entrepreneurship programs including investments
	Academic institutions	Contribute to incubation and acceleration programs (SMU, Singapore) Contribute to Innopreneurship-oriented academic programs (SMU, Singapore)
	Research Labs	*Inputs on market needs and wants both explicit and implicit* *Innovation brokerage*
	Large Enterprises	Active partnership to create new entities targeted for acquisition a la CISCO (USA) [CISCO]
	Investors	*Partner earlier and later stage investors for seamless progression of promising companies*
	Citizens	*Establish equity crowd funding for all stages of a company*
	Small- and Medium-sized Enterprises	*Actively advise on M&A opportunities* *Inputs on market needs and wants both explicit and implicit*

An innopreneurship platform's Triple Bottom Line score (IPTBLS) can be derived using the weights in Table 4.9.

In addition, i, a Reengineered Economy Score (RES) can be assigned to a smart city, where i is the number of economic transitions achieved by that smart city. i will take a value between 0 and n, where 0 represents a smart city that has not achieved a successful transition of its economy and n represents a city that had successfully achieved n transitions of economy.

Table 4.6. How citizens can contribute to other innopreneurship platform actors.

Contributor	Beneficiary	Components or programs
Citizens	Government	*Continuous feedback on social and environmental challenges*
	Academic institutions	*Inputs on demands for the workforce of the future*
	Research Labs	*Help validate demand for innovations*
	Large Enterprises	Feedback on current and future offerings Consumer inspired innovation requests (P&G, Zara)
	Investors	Micro-investments
	Citizens	Innopreneurship Mentorships
	Small and Medium Enterprises	Feedback on current and future offerings

Table 4.7. Values that SMEs can offer to other innopreneurship platform actors.

Contributor	Beneficiary	Components or programs
Small and Medium Enterprises	Government	*Demand for growth-oriented talent*
	Academic institutions	*Talent acquisition plans*
	Research Labs	*Know-how needs*
	Large Enterprises	Partnership, JV and M&A opportunities
	Investors	Growth oriented investment needs
	Citizens	Employment opportunities
	Small and Medium Enterprises	*Partnership to complete a value chain / network*

Table 4.8. An approach to scoring the links of other innopreneurship platform actors.

Value contributed to the beneficiary	Nature of value	Link score
Marginal	Positive	0.25
Average	Positive	0.5
Significant	Positive	0.75
Outstanding	Positive	1.0
None	Neutral	0
Marginal	Negative	–0.25
Average	Negative	–0.5
Significant	Negative	–0.75
Very significant	Negative	–1.0

Table 4.9. Triple bottom line scores for an innovation platform.

Nature of innovations created by an innovation platform	Innovation platform's TBL score
Any one of Economic, Social or Environmental dimensions	1
Any one pair of the above three dimensions	2
Any two pairs of the above three dimensions	3
All the above three dimensions	4
All innovations address all three dimensions	5

A Smart City Innovation Platform's score (SCIPS) can be computed using Formula 4.1.

$$\text{SCIPS} = \text{NIPS} * \text{IPTBLS} * \text{RES}. \tag{4.1}$$

The next section examines the role of key actors of a smart city's innopreneurship platform in its ability to regularly move its economy from one industry to the next.

4.5 Sustainable Smart City

A smart city can only qualify as a sustainable smart city if it is able to successfully move its economy from one generation to the next on a regular basis. A good example is the city of Lille in France, which has established an ICT incubator called Euratechnologies, by repurposing resources that had initially been used by the now-defunct textile industry [Euratechnologies]. Such sustainability can only be achieved if academic institutions and citizens are ready to respond in a timely manner to changes in technologies, markets, social developments and environmental concerns. The actions that can be taken by these key actors are discussed in the following subsections.

4.5.1 *Responsive academic institutions*

Academic institutions should ensure that they are able to infuse and sustain innovation culture across all their stakeholders. This can be achieved by defining policies and corresponding programs that are responsive to changing industrial landscape.

The academic institutions should also develop curriculum that nurtures and strengthens the key traits required to be an entrepreneur—traits such as Creativity, Exploration, Resilience, Experimentation, Adaptability and Learning to learn (CEREAL).

Where new industries may require a new set of skills, academic institutions should offer programs that help develop such new skills required by embryonic and emerging industries. Such efforts should include programs for those currently working in mature industries and are willing to learn new skills required by an emerging industry. Academic institutions should be aware of the lead time required to produce a skilled workforce for a new industry and hence have to carefully plan the launch of these programs to be in synchronization with the arrival of a new industry. This requires academic institutions to be a lot more agile than at present.

Academic institutions should offer Innopreneurship-related programs to all their stakeholders. An example of such programs is a progression of courses in innovation and entrepreneurship, beginning with identifying market needs and wants, and ending with building a prototype of an innovation that addresses those needs and wants. Other examples could include degree and non-degree programs for those already employed in current industries to transition into new industries, including programs that prepare individuals to become innovation managers and chief innovation officers.

Academic institutions tend to be the slowest to change. They often hang on to proven curriculum that has worked well. This is largely due to the fact that their faculty members themselves tend to focus on topics they are familiar with, continuing to work in areas where they have established themselves. For academic institutions to stay relevant, it is important to encourage and incentivize faculty members to evolve their research and teaching interests in alignment with emerging industrial trends and needs. It may also be important to provide disincentives for faculty members who continue to engage in topics that have become either outdated or irrelevant.

4.5.2 *Innovative citizenry*

A smart city can sustain its leadership only when a critical portion of its citizenry is Innopreneurs. Figure 4.6 presents the key traits of an Innopreneur.

Innopreneurs tend to be very curious. This curiosity gets them to learn how to learn about new topics. Their knowledge of new topics combined with curiosity eggs them on to explore new ideas for novel products, processes and services. Exploration is often followed by experimentation, to validate their new ideas. Experiments help validate aspects such as feasibility, affordability and scalability. Some ideas will succeed and others will fail. Innopreneurs tend to dust off their failures and use the learnings from their experiments to adapt their ideas to meet market demands. Such a process is often referred to as pivoting. A smart city cannot sustain its current leadership if their academic institutions do not produce a sufficient number of Innopreneurs.

All the citizenry of a smart city will have to accept the inevitability of change. It is well known that habits die hard and people hate change because change often makes their skillsets irrelevant and forces them to

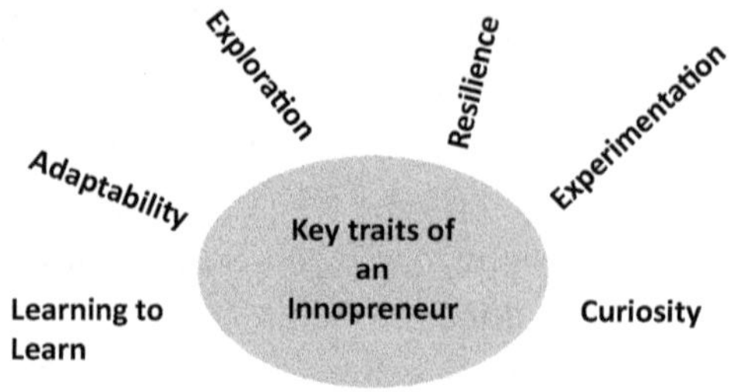

Figure 4.6. Key traits of an innopreneur.

compete with younger and more agile fellow citizens. Citizens who anticipate and embrace change by monitoring the sunset of their current industry and the rise of a new industry, and proactively acquire the skills required for the new industry become critical assets for the continuing leadership of a smart city. We call this "Minds Future"—a term that refers to a mindset that actively pursues and prepares individuals for the oncoming future. Such openness to changes can only be inculcated at a young age. It will be futile to expect those without such training in their younger years to transition from one industry to the next with ease in their later years.

4.5.3 *Roles of other actors*

Governments should and often do enact policies and programs for managing changes at least for the purposes of their own survival, if not for the betterment of their cities or countries. They have a choice of either pioneering such changes or quickly following others by studying how others successfully responded to changes, or choose to do nothing. Smart city governments that anticipate and effect changes in industries and erect infrastructures to support such changes will enhance the possibility of it retaining its leadership.

Research labs should engage citizenry to inspire them about the oncoming changes and explain the potential opportunities that will arise from embryonic and emerging industries. They should get individuals excited about preparing for the future through acquiring new knowledge and skills.

Businesses, both large and small, should invest in a city's citizenry to help them embrace and prepare for changes. Their contributions can be at multiple levels. They certainly can establish academic chair positions for faculty who create changes, offer scholarships to students pursuing innovative curriculum and support Innopreneurship programs of all kinds, including multi-tier competitions.

Investors should work closely with businesses to create companies for future industries. Investors such as venture capitalists should become "Venture Builders" who will partner companies in creating their future business divisions in emerging industries using the "Spinning-In" concept practiced by CISCO [Bort, 2014].

All actors should collectively create a "nation-wide/city-wide" Innopreneurship ecosystem learning from the tiered grooming as practiced by soccer academies, the classification system of International Chess federations classification, the weekly challenges of Hungarian Mathematics Academy, the apprenticeship practiced by Guilds and family involvement in little league games such as baseball.

4.6 Summary and Conclusions

This chapter has provided a partial list of examples of the policies, programs and contributions of actors in a smart city innopreneurship platform. This is hopefully a useful starter list for those aspiring to build smart cities. It has also examined the key roles that academic institutions and citizenry should play in both aspiring and established smart cities in order to build and maintain a smart city. It has listed the key traits of an Innopreneur and has suggested some methods for creating Innopreneurs.

Some of the established smart cities have done well due to a number of factors, key among them being a supportive government, responsive academic institutions and innovative citizenry. However, as these cities mature, they are bound to face a new set of challenges which may result in the migration of their Innopreneurs to greener pastures. Sustainable smart cities will be mindful of the naked truth—it is the innovative citizenry who can help maintain their leadership.

References

Bort, J. (2014). Why CISCO has showered three men with billions, *Business Insider*. URL: https://www.businessinsider.com/why-cisco-showered-three-men-with-billions-2014-9/?IR=T.

Brussels Smart City. *Brussels Smart City Project*. URL: https://smartcity.brussels/the-project.

Burggarf, H. (2018). How 'smart' is your city: Juniper Research reveals the 20 'Cleverest'. *International Investment*. URL: http://www.internationalinvestment.net/regions/asia/smart-city-juniper-research-reveals-20-cleverest/.

Cohen, B. (2014). *City in Transition: Urban Open Innovation Environments as a Radical Innovation*. Vienna.

EasyPark. (2017). Smartcity index https://easyparkgroup.com/smart-cities-index/.

Enterprise Singapore. Growing Enterprise. *Enterprise Singapore*. URL: https://www.enterprisesg.gov.sg/.

Euratechnologies. URL: https://en.euratechnologies.com/.

GOV.UK. (2013). What are Innovation Platforms? Innovation Platforms Practice Brief. 1, URL: https://assets.publishing.service.gov.uk/media/57a08a2840f0b652dd0005bc/Brief1.pdf.

IESE Business School. (2018). The smartest cities in the world in 2018. *Forbes*. URL: https://www.forbes.com/sites/iese/2018/07/13/the-smartest-cities-in-the-world-in-2018/#752799fe2efc.

Karolinska Development. *Who we are—We invest in Nordic Life Science Companies*. URL: https://www.karolinskadevelopment.com/en/who-we-are.

Karolinska Institute. *Welcome to KI Innovations—Innovation for a healthier future*! URL: https://karolinskainnovations.ki.se/en/.

Narasimhalu, A.D. (2012). *CUGAR: A Model for Open Innovation in Science and Technology Parks*. Daejeon, South Korea: WTA workshop.

SBIR. *Small Business Innovation Research*. URL: https://www.sbir.gov/about.

SBIR. *Israel–US Binational Industrial Research and Development*. Small Business.

Sophia Antipolis. *Invest in Cote D'azur*. URL: http://www.investincotedazur.com/en/sophia-antipolis/.

STTR. Small Business Technology Transfer. URL: https://www.sbir.gov/about.

Townsend, A. (2013). Smart Cities: Big Data, Civic Hackers, and the Quest for a New Utopia.

Warrier, P. (2018). Smart Cities need H.E.A.R.T. *Forbes*. URL: https://www.forbes.com/sites/forbestechcouncil/2018/04/13/smart-cities-need-h-e-a-r-t/#6669b221893d.

Wisconsin *Alumni Research Foundation*. URL: https://www.warf.org/.

Yozma. Creating a Professionally Managed Venture Capital Market in Israel. URL: http://www.yozma.com/overview/.

© 2023 World Scientific Publishing Company
https://doi.org/10.1142/9789811257186_0005

CHAPTER 5

Human-centric Smart City Development Strategy in Taiwan

Chien-Yuan Lin

National Taiwan University,
Taiwan Smart City Association, Taipei, Taiwan
Email: cylin@ntu.edu.tw

5.1 Introduction

Geographically, Taiwan is an island located in between Japan and the Philippines on the west of Pacific with a total land area of 36,193 square kilometers. By the end of 2019, there are 23.6 million inhabitants and about 70% of them live in the urban areas, and 15.3% of the total population were aged 65 and above. The GDP in 2019 is US$25,909 per capita [National Development Council, 2020].

Although there is no universally accepted standard definition of smart city so far, a smart city can be regarded as a city that undergoes development projects to integrate information and communication technology (ICT) and the Internet of Things (IoT) in a secure system to manage the assets of a city. In other words, any city can be a smart city once it applies urban informatics and technology to improve the efficiency of services and the quality of human life.

Similar to all other countries, Taiwan suffers from social problems of increasing aged population and decreasing birth rate. Facing problems of labor cost increasing, the development of a smart city is certainly a smart solution for social and economic problems. Taking advantage of the strong ICT industry development, the government administrations at both the central and local levels have promoted the development of smart city applications actively. In addition to the market forces, there are many other factors, including government support influencing the development progress. Based on the human-centric development strategy, by means of two-way (top-down and bottom-up) approach of project selection, various innovative applications have been incubated and developed for practical uses. In this chapter, the implementation strategy for the promotion of smart city development by both the government and private sectors in Taiwan will be introduced.

5.2 Digital Nation and Innovative Economic Development Program (DIGI+)

In order to promote the development of a smart city efficiently and effectively, a national policy to guide resource allocation is needed. To accelerate industrial innovation, and to create a so-called Smart Taiwan, the Taiwanese government has initiated a program named "Digital Nation & Innovative Economic Development Program (DIGI+) 2017–2025" in order to promote smart city development and to bridge the digital divide in rural areas. Each alphabet in the name DIGI+ has special policy implications. "D" stands for development, the first "I" stands for innovation, "G" stands for governance, and the second "I" stands for inclusion, and finally, the "+" sign indicates an "upgrade" in all areas.

The DIGI+ program is aimed to upgrade the infrastructure for broadband digital service, to provide friendly legal environment, to research and develop advanced digital technology, and to incubate inter-discipline digital talents. Furthermore, the program is to stimulate innovation and to develop the digital economy and trans-border electronic business, in order to support the development of a high value-added industry, and to establish digital government and a smart city.

Financial support is absolutely the key to promote smart city development. A total budget of about US$60 billion at the central government is allocated for DIGI+ to spend during 2017–2025. In order to promote and to coordinate the spending of this special budget across different ministries, a task force named "Digital Innovation &

Governance Initiative Committee" is established in the Executive Yuan directly supervised by the Premier. Within the task force of DIGI+, five divisions of administration units are established (DIGI+ Office, 2020).

The functions of each division are briefly described as follows:

(1) *Infrastructure Division*: To plan and develop the infrastructure for supra broadband cloud and digital flow service, and to develop a favorable environment for digital innovation, so that an equal and active network society can be created.
(2) *Technology and Talent Division*: To coordinate and promote the education and training of multidisciplinary talents for digital innovation, to research and develop advanced digital technology, and to plan international collaboration strategies and measures.
(3) *Digital Economy Division*: To establish digital innovation ecosystem, to accumulate digital innovation capability, and to support trans-industrial transition and upgrading, to enhance the cross-department digital economy, a healthy environment for digital service industry and international development.
(4) *Digital Nation Division*: To create a friendly legal system and a safe and reliable application environment, to provide open data and service-based digital government.
(5) *Coordination and Promotion Division*: To monitor the overall development status and trends, to plan and initiate related strategies and measures, to coordinate ministries and local governments for actions, and by means of the platform of industry–public–university collaboration, sustainable and smart cities are to be developed.

5.3 Two-Way Approach with Public–Private Partnership

Taking advantage of the ICT manufacturing industry, the Taiwanese government has actively promoted the development of smart city applications in many ways. Among all local cities, Taipei city is the first one developing ICT applications to improve public service efficiency and quality, in the name of an intelligent city.

Figure 5.1. Evolution of smart city development strategy.
Source: DIGI+ Office [2020].

Figure 5.1 shows that the strategy of smart city development in Taiwan has evolved in three stages. In the first stage before 2008, the so-called e-Taiwan project was promoted by the central government with focus on the transformation of E-government. Essentially, it was a technology-driven strategy where system development very much relied on vendors' solutions. In the second stage (2008–2017), the local government began to actively lead the development of ICT infrastructures and smart city applications. In fact, Taipei city was invited to demonstrate its smart city applications in the 2010 Smart City Summit & Expo (SCSE) in Shanghai, China. Meanwhile, the central government focused on the expansion and upgrade of the mobile telecommunication system for smart city solutions.

Figure 5.1 shows that public–private partnership model was introduced into the smart city development strategy in 2018. In order to be more responsive to local demands and market feasibility, based on the public–private partnership model, the two-way (Top-down and Bottom-up) approach for project selection was adopted in Stage 3. To select appropriate projects to support candidate company proposals are firstly submitted to the government for test areas identified, go through a review process, and finally the winners will gain a certain amount of financial support from the government.

Figure 5.2 shows the operation process of the two-way approach (top-down and bottom-up) in the promotion of smart city developments. The total budget is divided into two parts; (1) one is to fund

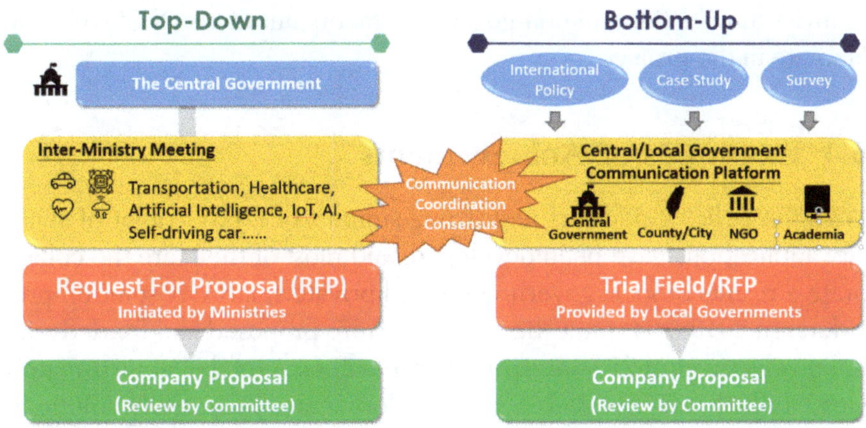

Figure 5.2. Smart city promotion with two-way approach.
Source: Industry Development Bureau [2019].

company proposals by the top-down approach and (2) the other part is to fund company proposals by the bottom-up approach. The top-down approach is actually the traditional way of funding allocation that ministries at the central government decided the application areas firstly, and then documents of request for proposal (RFP) are prepared and open to the public, and then company proposals from solution providers are submitted for review and selection. The advantage of the top-down approach is that resource allocation can be sure to be policy responsive at the national level. However, it may not be able to meet local demands and there can be a lack of financial feasibility for the final product.

By means of the bottom-up approach, smart city solution providers are required to collaborate with local governments in order to submit company proposals for the competition of funding support. With the collaboration, the local government will help the solution providers to conduct field trial in specified places; meanwhile, the solution providers need to share about half of the total expense in the system development. Furthermore, in order to receive the total funding amount at the end of the project, the company has to prove that the final product functions well on field trials and it reaches the performance requirements pre-set in the proposal. Certainly, the smart city solution providers are encouraged to invite universities and non-governmental organizations (NGOs) to collaborate in the project.

5.4 Examples of Achievements

With the policy guidelines and strong support of DIGI+, a lot of smart city applications have been developed, and most of them are the collaborative results of local governments and private smart city solution providers. In 2019, 164 innovation projects were proposed and funded with the involvement of 223 enterprise participators. A total of approximately US$1 billion was invested in the development of smart city solutions. As a matter of fact, because of the significant performance resulted from innovation projects, many local governments had won international awards. For instance, the smart street light management system in Taoyuan City won the first prize in the 2019 Intelligent Community Forum (ICF) competition. Five interesting cases with significant benefits in different smart application areas are briefly introduced as follows.

Case 1: Smart Parking Service with IOT

Parking difficulty is a headache to almost every city. Poor parking service not only wastes drivers' time in space searching, but also causes unnecessary trips to jam the street traffic. A lot of cities in the world work hard to improve their parking management system by means of smart city solutions. The smart parking service system developed in Taipei is one of the leading applications of IOT technology [Taipei City Government, 2019].

In the test area, more than 40,000 geomagnetic sensors were installed under the pavement to detect empty parking spaces and parking time. Meanwhile, the system is integrated with smart signage on the roadside, Google Maps, and mobile payment in order to provide a one-stop parking services for the vehicle drivers. The smart parking service system can provide real-time information about parking space occupancy, direction and diverse payment for drivers, so that searching time can be significantly reduced. The result of the pilot project showed that an average of six minutes can be reduced for every parking vehicle. Moreover, the turnover rate of parking spaces is also improved by 5%.

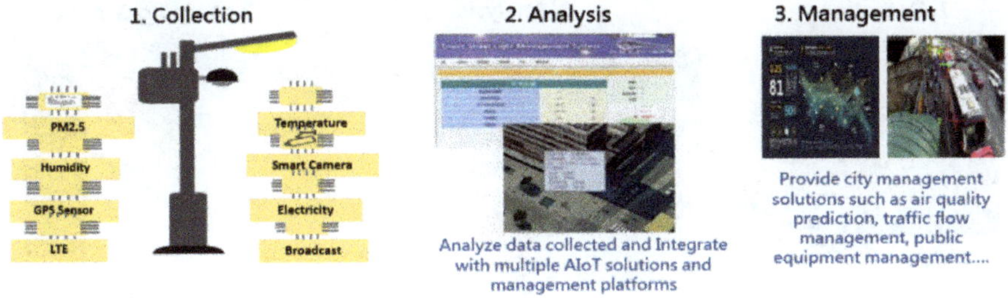

Figure 5.3. Smart street light system.
Source: Industry Development Bureau [2019].

Case 2: Smart Street Light for city management

Currently, more than 7,500 smart street lights have been installed and equipped with multiple IoT sensors in Taoyuan City, Taipei City, Chiayi City and Keelung city. Street lights and various sensors are integrated with AIoT (AI + IOT) and Multi-access Edge Computing (MEC) to establish an integrated platform to facilitate the street light management for the local governments.

Taking advantage of the street light poles along the roadsides, many other sensors are installed to collect data related to air quality, temperature, and traffic detectors. As shown in Figure 5.3, not only the lighting electricity can be managed more efficiently, air quality, road traffic and public facilities can also be managed more effectively [Taoyuan City Government, 2019].

Case 3: Monitoring Air Quality with AI Tech.

Most areas near manufacturing factories often suffered from air pollution, and this is especially serious for areas with petrochemical industries. Although regulations already existed, enforcement is very difficult since it is difficult to detect the source of emission. In Taoyuan, an AI-based air pollution monitoring platform is developed to help the industrial park management office to detect air pollution sources, to monitor potential sources, and to conduct inspections or to take response measures immediately. Figure 5.4 shows the framework of the AI-based air quality monitoring system.

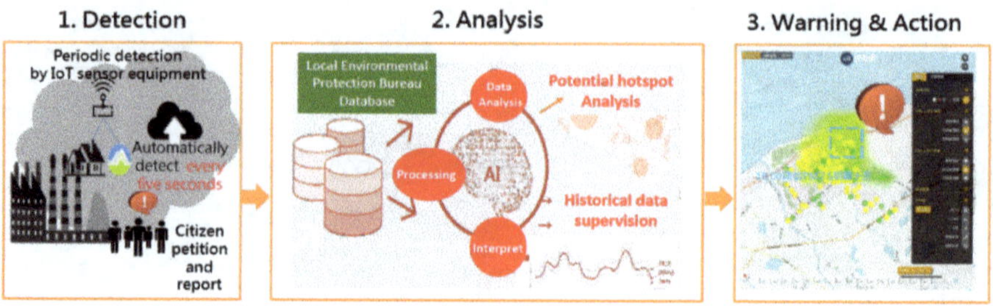

Figure 5.4. AI-based air quality monitoring system.
Source: Industry Development Bureau [2019].

By use of the AI algorithm, the system can predict air quality changes four hours in advance. The experimental project also shows that it can help local environmental protection agencies to improve investigation efficiency by six times and reporting rate by 70%.

Case 4: Safeguard for Elderly with Alzheimer's Disease

The increasing number of aged populations has caused various health and care problems. One of the difficulties is to take care of elderly with Alzheimer's disease. Although families hope their elderly can go outdoors to enjoy their life, they fear that demented elderly gets lost and it is not easy to find back. Therefore, Pingtung County cooperated with China Telecom to develop a safeguard system for the elderly by the use of IoT technology. The so-called lucky charms were developed and delivered to elderly and equipped with beacon sensors, in order to provide positioning service via IoT network. Figure 5.5 shows the functional components in the safeguard system with IoT technology.

Case 5: UAV solutions for smart agriculture

Along with economic development and urbanization, a lot of young people are attracted to migrate to the cities from the rural areas. As a result, it leads to farm labor shortage. As shown in Figure 5.6, with the unmanned aerial vehicle (UAV) smart agriculture solution, farmers can improve their working productivity by increasing the spraying efficiency

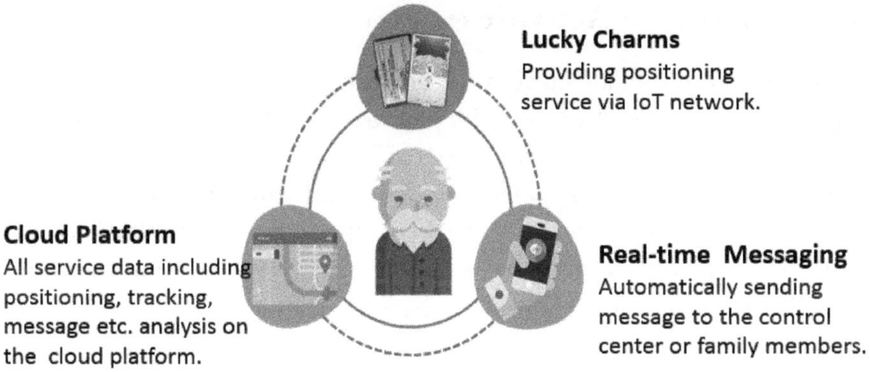

Figure 5.5. Lucky charm for elderly safeguard.
Source: Industry Development Bureau [2020].

Figure 5.6. UAV solutions for smart agriculture.
Source: Industry Development Bureau [2020].

for more than 40 times; meanwhile, the risk of poison exposure can be reduced.

With the funding support of the Industry Development Bureau, Ministry of Economic Development, Jingwei Aerospace Technology had developed the UAV-based smart agriculture solution for agriculture production. To make the application more financially and technically feasible for farmers, the UAV hardware company has transformed itself to became a UAV service rental company. Through the technical service verification on farmlands in Taiwan, aerial application standards are established and the product system is currently promoted to overseas markets [Industry Development Bureau, 2019].

5.5 Supporting Activities

In addition to the financial support provided by the government sector, supporting activities offered by the private sector have also made significant contributions to the promotion of a smart city. Those supporting activities provide non-financial incentives and stimulations to promote the development of innovative smart city applications in both the public and private sectors. Among various supporting activities, four important activities are introduced as follows.

5.5.1 *Smart city EXPO and competition*

Both conferences and expositions are important interchange platforms for quick and effective communication among firms and people who are interested in and are related to smart city products and services. In fact, the explosion provides an opportunity for people to learn within a short time and to gather people for business match-up and expertise interaction. The first international Smart City Summit & Expo (SCSE) in Taiwan was sponsored and held by the Taipei Computer Association (TCA) in 2014. The TCA is the leading industrial organization in Taiwan with more than 4,000 members from various fields such as software, hardware, semiconductors and components, manufacture, sales, network communication service, etc., thereby generating over 80% of the total production value of Taiwan's ICT industry.

After six years, the SCSE in Taipei has gradually become an internationally famous conference and exposition on the smart city and IOT industry. In fact, the SCSE is now the largest smart city tradeshow in the Asia-Pacific region. In the annual SCSE, three types (G2G, B2G, and B2B) of functional activities are arranged. For instance, the smart city summit has attracted mayors and delegates from more than 120 cities to share experiences on government-to-government (G2G) basis. By means of forums, exposition and seminars on various topics, B2G and B2B interchanges are conducted. In 2019, 337 exhibitors (with 1,250 booths) attended and 33,216 professional visitors were attracted to the activity. There were 128 city leaders from 45 countries who attended the Mayor's Summit, while 17 city/country pavilions attended in the Expo. Furthermore, it is a platform for cross-industry business collaboration. Meanwhile, 66 forum tracks were conducted for various areas, including smart city forums and the World Telecom Smart City Conference. Unfortunately, due to the impact of COVID-19, the 2020 SCSE was postponed to July 2020 and conducted online [Taipei Computer Association, 2020].

In parallel to the SCSE, competitions such as "Awards of Smart City Application" is conducted by the TCA annually. The competition is composed of two groups, namely, government groups and business groups. Every year, more than 100 cases in each group attend the competition, and only those who are very outstanding and impressive cases can receive the awards and they are honored at the opening ceremony of the SCSE.

5.5.2 *City project management office*

In addition to the central government, the local government has played a key role in the promotion of smart city development. In the implementation of DIGI+ program, the Council of National Development has funded almost every local government to prepare a comprehensive plan for the development of a smart city and to establish a top-level office to coordinate and to promote the implementation. Among all local governments, Taipei city was the first one establishing a specialized task office to promote smart city development with significant achievement. In 2016, the Department of Information Technology, Taipei City government established the "Taipei Smart City Project Management Office", also known as the "PMO", to implement and supervise the development of a smart city. The office also functions as an innovation matchmaking platform to combine industry and government resources to develop smart solutions to satisfy public demands.

The goal of the PMO is to promote public participation and public–private partnership in order to create new technologies, innovative applications, and data/information to solve citizens' problems. The Taipei city government hoped to construct an ecosystem so that the government, the industry, and citizens can share the prosperity from its development. Furthermore, it was planned to establish a matchmaking platform which allows citizens access necessary innovative technologies.

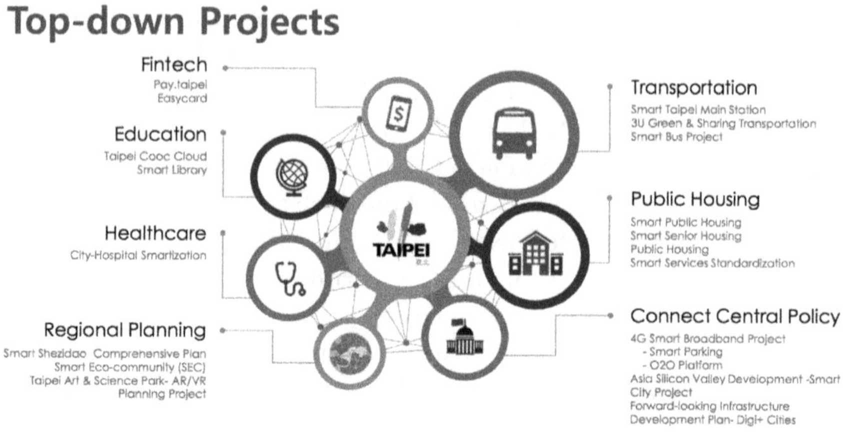

Figure 5.7. Taipei smart city as living lab.
Source: Taipei Smart City Project Management Office [2019].

It will gradually open up many fields allowing interested industry to do trials and make the city to be a so-called "living lab". Figure 5.7 shows the application areas of the living lab as promoted by the Taipei Smart City Project Management Office [2019].

5.5.3 *Annual survey of smart city development status*

In the promotion of smart city development in Taiwan, public media also plays an important role. With the collaboration of *Wealth Magazine*, the Taiwan Smart City Association (TSCA), and Unique Satellite TV, an annual survey of smart city development status was initiated in 2016. The most recent survey was completed in July 2020. The purpose of the survey is to evaluate the scope and level of smart city applications in all local cities. Results are ranked and introduced to the general public through special reports and programs of *Wealth Magazine* and Unique Satellite TV. By so doing, the general public can be better informed and mayors of local governments can be better responsive to what the people care about. Survey areas include smart governance, smart healthcare, smart transportation, smart energy, smart security, and smart buildings.

Before the survey, an expert committee is established to help the preparation of the questionnaire and to review and to help with the statistical analysis after the survey. Two sets of questions were surveyed at the same period. One survey is aimed at the sampled general public with telephone interviews, and in the other survey, questionnaires are sent to each local government to be filled in by their designated officials. For the general public version, through sampled telephone interviews, people are asked to reveal their willingness to use the proposed application and the frequency of practical use. For the government version, all local governments are surveyed and each designated a representative to report their plans and progress of system development.

The survey data are analyzed in three ways. The first part is to compare the progress and performances among all the local governments. Results in every year show that all metro cities perform much better than all other non-metro cities. The second part of the analysis is to compare performances with previous surveys. Taipei City used to lead all cities on all areas since 2016; yet, the 2020 survey shows that Taipei City can only lead in a few areas. Obviously, cities are competing with each other and working very hard to make progress. The third part of the analysis is to check the progress of different application areas. Taking the 2020 survey as an example, it is found that more than 90% of sampled respondents support the government to apply smart city technology in its fight against COVID-19, even though privacy may be sacrificed [Sun, 2020].

5.5.4 *Awards of excellent contribution to smart city development*

People are the key in promoting the development of smart city applications in the real world. Instead of recognizing the achievements of an organization or a product, the "Awards of Excellent Contribution to Smart City Development" is to recognize and honor an individual's excellent contributions to the development of a smart city. The "Awards of Excellent Contribution to Smart City Development" competition was initiated in 2018 and it is sponsored by the Taiwan Smart City Association (TSCA), which is a non-profit organization. The competition covers 10 areas, namely, smart governance, smart community, smart medical care, smart security, smart transportation, smart tourism, smart energy, smart agriculture, smart service and smart innovation. Participants should submit their applications with an introduction of their contributions and recommendation letters by the end of June each year.

Winners are to be honored with a Certificate of Award and a crystal plaque in a formal ceremony. From 2018 to 2020, a total of 22 winners have been awarded in 10 areas. Every year at the awards ceremony, high-ranking government officials are invited to hand the award plaques to the winners. Through TV broadcasting and news report of public media, the winners are introduced and honored to the public. Figure 5.8 is the group photo of winners with Vice President Chin-Der Lai and other official VIPs in the 2020 award ceremony. Among all winners, it may be noticed that the winners of smart governance in 2019 and 2020 are the Mayors from Taoyuan City and New Taipei City respectively; actually, these two cities have won the ICF awards for many times in

Figure 5.8. Ceremony of 2020 "awards of excellent contribution to smart city development".
Source: Taiwan Smart City Association [2020].

recent years. It is evident that winners of the "Awards of Excellent Contribution to Smart City Development" are highly honored.

5.6 Summary

The purpose of smart city development is to use smart technology to bring benefits of a better life, better economy and better environment to the people in a more sustainable way. Due to the limitation of resources, every government needs to find an efficient and effective way to promote the development of a smart city. In other words, the challenge to the government is not the procurement of the most expensive ICT equipment; instead, the challenge is how to allocate limited resources to maximize the benefits for the identified application areas with the appropriate smart technology. To share the experiences of smart city development, the human-centric strategy being adopted by the Taiwanese government is presented in this chapter.

Based on public–private partnership and by means of two-way (top-down and bottom-up) approach in the project selection, it is evident that the DIGI+ program based on both approaches has created significant contributions to the development of smart city application. The supporting activities sponsored by different private organizations also support the development of a smart city indirectly. However, just like all other public policies, implementation strategies need to be flexible and responsive to accommodate the changing socioeconomic environment. Looking into future, three policy directions are suggested to further the development of smart city applications in Taiwan.

(1) It is evident that the bottom-up approach has indeed encouraged the local governments to collaborate with different ICT firms to develop various applications for local demands. However, some application systems developed by different local governments are aiming at the similar service provision, but with different system specification. In other words, once a city has collaboratively developed an application system with an ICT firm, then the city can only rely on that particular firm for system operation and maintenance, and systems cannot be easily transplanted to different cities.

For long-term benefits, it is necessary for the government to standardize the system specification in the future.

(2) With the bottom-up approach, most of the application systems are developed independently to focus on a specific service area. In order to create a synergy effect, related application systems should be integrated in the future. For instance, smart parking management system should be integrated with the smart area traffic management system so that a smart transportation system can be more effective for the wide area.

(3) The disadvantage of funding incentive for public–private partnerships (PPP) projects proposed by private firms is that only projects with market potential can attract them to join. In other words, those socially and economically disadvantaged groups and areas can hardly attract the attention of system development. It is the responsibility of the public sector to provide the necessary public services for the disadvantaged groups, even though the system does not generate any profit.

References

DIGI+ Office. (2020). URL: https://digi.ey.gov.tw/Page/A1701D7654C6B83C.

Industry Development Bureau. (2020). URL: https://idb.org.tw/?p=about.

National Development Council. (2020). URL: https://www.ndc.gov.tw/default.aspx.

SCSE. (2020). URL: https://smartcityonline.org.tw/?p=about.

Sun, J.P. (2020). 2020 Survey of smart city development status, *Wealth Magazine*, *614*, 124–142.

Taipei City Government. (2020). URL: https://smartcity.taipei/.

Taipei Computer Association. (2020). URL: https://www.tca.org.tw/en/.

Taipei Smart City Project Management Office. (2019). URL: https://smartcity.taipei/posts/53.

Taiwan Smart City Association. (2020). URL: https://tsca.org.tw/?p=about.

Taoyuan City Government. (2019). URL: https://www.tycg.gov.tw/eng/.

© 2023 World Scientific Publishing Company
https://doi.org/10.1142/9789811257186_0006

Chapter 6

Did Users Co-create Innovation in Smart City Initiatives? A Case in Japan

Jun Yamashita

Social and Cultural Studies,
Kyushu University, Fukuoka, Japan
Email: yamasita@scs.kyushu-u.ac.jp

6.1 Introduction

Since Gibson *et al.* [1992] coined the term 'smart city', theories and practices of smart cities have flourished. Regarding the theoretical aspect, definitions of a smart city have become more diverse as the number of smart city developments has flourished [Albino *et al.*, 2015; Lazaroiu and Roscia, 2012]. Among such diverse definitions, Caragliu *et al.* [2009] considered "a city to be smart when investments in human and social capital and traditional (transport) and modern (ICT) communication infrastructure fuel sustainable economic growth and a high quality of life, with a wise management of natural resources, through participatory governance" [p. 50]. According to this definition, participatory governance, including citizens' participation, is regarded as a part of the smart city concept.

Following the aforementioned definition, theories on user-driven innovation, which not only leads to sustainable economic growth and a high quality of life but is also a part of the smart city concept, have been developed within the framework of smart city innovation ecosystems. While defining a smart city as "environments of open and user-driven innovation for experimenting and validating" internet services, Schaffers *et al.* [2011a] emphasized that the establishment of such innovation ecosystems "requires sustainable partnerships and cooperation strategies among the main stakeholders" including users [p. 431]. Following Schaffers *et al.* [2011a], Komninos *et al.* [2013] pointed out the importance of co-creation by both producers and users, and characterized smart cities as having "a high level of citizen involvement in co-creating Internet-based applications in all sectors of the economy and society" [p. 120]. Besides, Jorna and Veenstra [2015] specified the following three components of innovation ecosystems, which enable both top-down and bottom-up development of smart cities, namely a user-driven approach—(1) an open infrastructure, (2) needs-based value creation and (3) multi-party open governance. The most comprehensive framework regarding innovation ecosystems for smart cities may be the one presented by Zygiaris [2013], the smart city reference model. This model consists of six layers following the basic layer (Layer 0) community involvement, from the bottom (Layer 1) to the top (Layer 6): the green city, interconnection, instrumentation, open integration, application, and innovation layers. Urban planners might utilize this reference model to define smart city planning concepts and characterize smart city innovations.

Partly based on the aforementioned theories of smart city innovation ecosystems, case studies on user-driven innovation have been undertaken. These studies have mainly focused on the possibility of integrating open innovation platforms, namely living labs, with internet applications in various areas. Schaffers *et al.* [2011b] proposed an integrated system of living labs with internet access under three European Union (EU) FP7-ICT projects, and, to a degree, verified the validity of this system, while Komninos [2011] showed that the combination of user-driven innovation and ICT made smart cities more efficient in city governance, in the cases of Bletchley Park in the UK, Hong Kong, and Amsterdam in the Netherlands. The results of a case study undertaken in Vancouver, Canada by Lea *et al.* [2015], revealed that "community-driven ecosystems are more robust and self-sustaining", whereas, ordinary "innovation ecosystems are fragile and need nurturing" [p. 1,542]. However, it is possible to argue that, to date, the research on smart cities based on user-driven innovation has limitations. Besides, these empirical studies were chiefly conducted in European cities with only a few studies being undertaken in other parts of the world. This indicates that further smart city research is necessary, especially in a non-European context. The present chapter, therefore, examines user-driven innovation in smart city developments that took place in Japan.

Regarding the practical aspect of smart cities, various projects and programs have been, and are currently being, implemented globally. A wide range of smart city programs established by the Digital Agenda for Europe, commencing in the EU, in 2010, is an example of such smart city policies. Regarding smart city policies in Japan, it is clear that Japan has also experienced similar global trends in smart cities since 2010. The Japanese National Smart City Policy was introduced in 2010. At that time, the Agency for Natural Resources and Energy, under the Ministry of Economy, Trade, and Industry, initiated a national smart city policy entitled the Next Generation Energy and Social System Demonstration Project (hereafter, the smart community project, for short). Four cities were selected to participate in this smart community project. Another national smart city policy implemented in 2011 was the Eco-Model and Future Cities Construction Project, managed by the Cabinet Office on the basis of the Future City Initiative. In the future city project, 23 cities were designated as eco-model cities, and 11 cities were designated as future cities. Although both smart city national policies aimed for the construction of smart sustainable cities, the directions of the smart community and future cities projects were different. While the former was closely related to the term "smart", namely ICT, the latter was oriented toward environmental sustainability.

Simultaneously, some Japanese companies have transferred their smart city-related technologies overseas since 2010. The Becamex Tokyu, a joint venture between Tokyu Corporation and Becamex IDC Corporation in Vietnam, launched the Tokyu Binh Duong Garden City Project in Binh Duong province, Vietnam, in 2012. In 2013, construction of a mass rapid transit (MRT) began in Jakarta, Indonesia, by a joint venture between Shimizu Corporation and some Indonesian companies. Since the end of these smart community projects in 2014, the experiences have resulted in some technologies being transferred to both domestic and international markets.

6.2 How to Evaluate Co-creation of Innovation by Users?

The research subjects in this chapter were the four cities, Yokohama, Toyota, Keihanna, and Kitakyushu, selected for the aforementioned Smart Community Project run by the Agency for Natural Resources and Energy from 2010 to 2014. A smart community was defined as "a new form of social system that comprehensively manages the supply and demand of energy in the distributed energy systems, optimizes the use and application of energy, and incorporates lifestyle support services including monitoring services for the elderly, through the energy management system utilizing information technology (IT) and storage energy technologies, while making use of distributed energy resources such as renewable energy and cogeneration" [Japan Smart Community Alliance, 2015, p. 2]. A smart community was also regarded as a community where there was an "effort to change the social system of a defined area into a smarter state with technologies not only for electric power systems but also for a variety of public infrastructures including heat supply, water and sewerage, transportation, and communications" [*Ibid*]. The concept of such a smart community is presented in Figure 6.1.

The reason for the smart community project was the intention of the Japanese government to change the energy supply system in Japan. Since the end of World War Two, 10 electric companies have oligopolized the supply of electricity in 10 large regional blocks. Furthermore, Japan is a country prone to natural disasters. After Japan experienced large-scale disasters, especially the 2011 Tohoku Earthquake and Tsunami, decentralizing the supply of power and introducing flexible power pricing at the local level has been emphasized in order to avoid large-scale electricity shortages, thereby bringing an end to the oligopolized electricity supply. Renewable energy resources have also become regionalized so that these resources may be incorporated into local energy supplies in smart communities using ICTs.

In the present chapter, the Quadruple Helix Model was employed as the analytical framework for the user-driven innovation. The Quadruple Helix Model, based on a collaboration between industry,

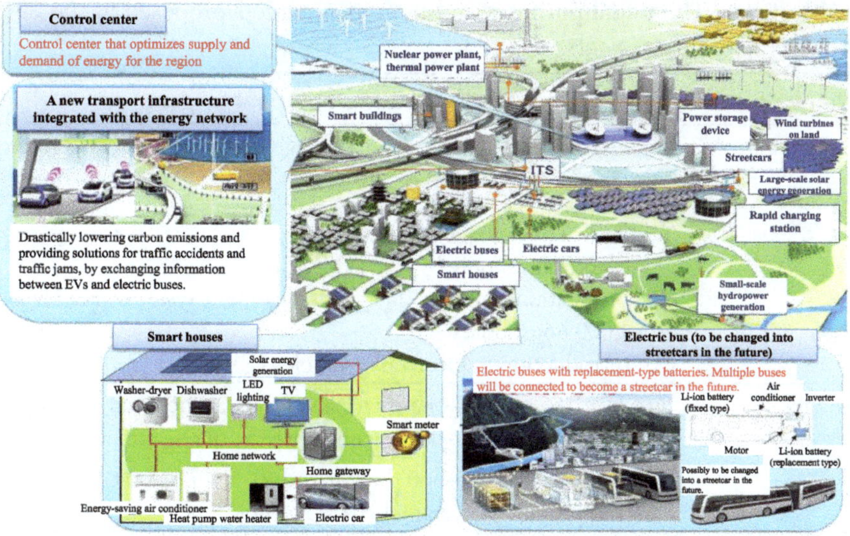

Figure 6.1. The concept of a smart community.
Source: Ministry of Economy, Trade, and Industry [2017].

government, academia, and society [Carayannis and Campbell, 2009], was regarded as the theoretical basis for the two aforementioned Japanese national smart city policies because these policies targeted open innovation [Bibri and Krogstie, 2017]. Regarding applications of the Quadruple Helix Model for smart city development, Martinez *et al.* [2016] discussed relationships between smart cities and the Quadruple Helix Model in the context of healthcare. They pointed out the necessity of inclusive spaces, namely living labs, as a policy tool of the Quadruple Helix Model. Such spaces comprising of health services and end-users such as citizens, patients, relatives, healthcare workers, and other professionals foster inclusive innovation, such as welfare technology, eHealth, telemedicine, and mobile health solutions. Mora *et al.* [2018], moreover, discussed the superiority of the Triple or Quadruple Helix Model regarding smart city development involving four European cities. However, smart city studies using the Quadruple Helix Model are limited; therefore, further studies are also necessary. Thus, the present chapter intended to address such limitations.

Table 6.1. The initial purposes and numerical targets proposed in the 2010 smart community projects.

Name of city	Purposes and numerical targets
Yokohama	*Purpose*: To employ an energy management system in an existing large metropolitan area. *Numerical targets*: (1) Electricity supply by PV: 27MW. (2) Installation of home energy management systems (HEMSs): 4,000 households. (3) Introduction of electronic vehicles (EV): 2,000 vehicles.
Toyota	*Purpose*: To optimize energy usage and traffic demand and supply in the household sector in consideration of consumer behavior. *Numerical target:* Carbon dioxide reduction of 8,000 t-CO_2/year in the project area, namely 67 households.
Keihanna	*Purpose*: To build a city with the lowest amount of carbon dioxide emissions per unit for all energy consumption in Japan without sacrificing quality of life (QOL) standards. *Numerical target*: The carbon dioxide emissions per capita per year should be 1.6 t-CO_2/person/year in the project area.
Kitakyushu	*Purposes*: (1) To transform ordinary energy consumers, namely residents, business establishments and others, into production consumers (prosumers) via the installation of photovoltaic power generation panels, etc. (2) To realize "demand-side self-management" by enabling prosumers to manage their own energy usage. (3) To introduce a combined mechanism of dynamic pricing and incentive programs. *Numerical targets*: (1) Carbon dioxide reduction: A 50% cut in the project area compared with regular city blocks. (2) Stable electricity supply: To maintain frequency and voltage fluctuation within a certain range (a voltage of 101 ± 6V, a frequency at 60Hz) even when introducing a large amount of new energy.

Source: Ministry of Economy, Trade, and Industry [2011].

Using the final reports of the four smart community projects, the involvement of the four subjects in the Quadruple Helix Model, namely the government, academia, industry, and society, were examined to identify user-driven innovation in smart cities. Along with examining innovation ecosystems, the present study evaluated the achievements of each smart community project. Table 6.1 summarizes the initial goals and numerical targets proposed in each project in 2010.

6.3 Did Users Co-create Innovation?

In this section, firstly, project outlines are described, and then the outcomes are evaluated. Finally, the validity of the Quadruple Helix Model is examined.

6.3.1 *Project outlines*

The results of each smart community project are summarized in Table 6.2.

6.3.1.1 *Yokohama city*

The Yokohama Smart City Project (YSCP) was implemented in an existing metropolitan area. Yokohama City has a population of approximately four million. In this project, around 4,000 households had photovoltaic (PV) panels and other smart equipment installed. This equipment was managed by home energy management systems (HEMSs) (see Table 6.3). Building energy management systems (BEMSs) were also established in 10 buildings. These HEMSs and BEMSs were connected to community energy management systems (CEMSs) to control peak cut and in turn conserve energy in order to reduce carbon dioxide emissions at the regional level.

6.3.1.2 *Toyota city*

Toyota City, which has a population of less than 420,000, is famous for being the location of Toyota Motor Corporation's headquarters. This

Table 6.2. A summary of the results of the four smart community projects.

Yokohama City (Wide-area metropolis) • 4,000 households and HEMSs. • 10 large-scale buildings and BEMSs. • Multiple storage batteries.
Toyota City (Separate housing) • Local production for local consumption. • 67 households equipped with solar panels, household fuel cells, and storage batteries. • Advanced transportation system (EV, PHV).
Keihanna City (Housing complexes) • 700 households and HEMSs. • Consulting businesses regarding energy conservation.
Kitakyushu City (Designated supply area) • Power is supplied by Nippon Steel & Sumitomo Metal Corporation. • Dynamic pricing system for 180 households.

Source: From Tobe, C. (2015), *Japan's policy on Smart Communities*.

Table 6.3. Summary of the Yokohama smart community project results.

HEMS • Solar power generation, fuel cells, solar heat, EV, etc. were introduced into existing housing complexes. • Energy usage fell by approximately 40% through the utilization of renewable energy and distributed energy, interchange of electrical heat/integrated control, and the introduction of HEMSs. • An 80% or higher rate of self-sufficiency in electrical power energy.
Integrated BEMS • Two buildings were managed using an "integrated BEMS" and such BEMSs were also connected with each other. • Approximately 10% more energy was saved compared to ordinary energy-saving buildings.
CEMS • Large-scale demand responses were implemented.

Source: Komiyama [2012].

Table 6.4. Summary of the Toyota smart community project results.

Energy Management
• There are 67 homes equipped with solar panels, household fuel cells, Eco Cute, secondary cells, plug-in hybrid vehicles, electric vehicles, etc., were newly introduced in an urban district of Toyota City.
• From the perspectives of local prosumers and cutting back during peak hours, pseudo-dynamic pricing was implemented, and digital currency was awarded through smartphones, etc. HEMSs were also linked to an EDMS (energy data management system) to manage energy for the entire district.
• Plug-in hybrid vehicles (PHV) and electric vehicles (EV) were used as secondary cells.
Transportation System
• The driving conditions of PHVs and EVs were monitored using the EDMS. Using this system, optimal driving routes to destinations were presented for each vehicle in accordance with road conditions, and point-based incentives were offered as rewards to energy-efficient drivers.
• Power supplies for fuel cell buses and schools, among others, were demonstrated in the event of natural disasters.

Source: Komiyama [2012].

company took a leading role in the Toyota City Low-Carbon Society Verification Project (Smart Merit). Unlike the Yokohama Smart City Project, the project in Toyota City was implemented in a small town named "Toyota Ecoful Town". This housing complex consisted of 67 newly-built houses installed with HEMSs. Along with the energy management in houses, this smart community project also focused on the transportation sector. Privately-owned electric and fuel cell vehicles were provided with the aforementioned HEMSs as storages and generators (see Table 6.4).

6.3.1.3 *Keihanna city*

The smart community project entitled the "Keihanna Eco City Next-Generation Energy and Social System" was run by a local government consortium under the auspices of an academia–industry–society collaboration (see Table 6.5). This is partly because Keihanna City consists of eight municipalities spanning three prefectures with a population

Table 6.5. Summary of the Keihanna city smart community project results.

HEMS (for 14 homes)
• HEMSs measured energy consumption by electrical apparatus. • HEMSs were also utilized as energy and storage battery management systems.
BEMS (for the Keihanna Plaza)
• Reduction of CO_2 emissions through fuel optimization and suppression. • DR (demand response) measures were taken for not only approximately 70 building tenants and common spaces but also hotel rooms.
CEMS
• Identified the status of regional energy demand and supply, including energy and gas usage statistics for households, buildings, and EVs in the region, as well as PV energy generation. • In order to draft an optimal energy usage plan, each EMS was developed and the effects of DR and balancing measures were verified.
Large-scale Demand Response (for approximately 700 homes)
• The energy demand suppression effect of price-induction was verified for approximately 700 households. • Effects of energy-saving consultations based on energy-usage data were also verified.
EV Charging Management (for approximately 100 vehicles)
• This system collected and managed EV position, battery levels, and operation data. • The effects of changing and energy suppression in harmony with the CEMS were also verified.

Source: From Shigematsu, C. (2015). *Creation of Smart Communities in Keihanna Science City—From Demonstration to Implementation.*

of approximately 240,000. Among these local governments, Kyoto Prefecture and three municipalities under this prefecture participated in this smart community project. Likewise, the Yokohama Smart City Project, the smart community project in Keihanna, comprised approximately 700 HEMSs with around 100 EVs, one BEMS and one CEMS to manage demand response and peak cut.

6.3.1.4 *Kitakyushu city*

This city, with a population of under one million, has been recognized as one of the world's most sustainable cities [OECD, 2013]. The Kitakyushu Smart Community Project focused mainly on hydrogen (see Table 6.6). Houses in the project area were connected to a hydrogen factory using pipelines. These houses generated electricity using their own fuel cells, and sometimes electricity was stored as hydrogen. Using such innovations, this project also aimed to change electricity consumers to production-consumers known as "prosumers".

As a common characteristic of the four smart community projects, the EMS was the main pillar of these projects, though some projects were related to the transportation sector, mainly concerning privately owned vehicles. This characteristic may result from the fact that the Agency for Natural Resources and Energy managed the smart community policy. Therefore, the use of renewable energy was limited to PV and wind power. The utilization of biomass was not taken into account. For these reasons, neither the waste sector (including biomass rich sewage systems) nor the transportation sector (including public transportation systems using biogas generated from these sewage systems) was integrated into the Japanese smart community development program. The aforementioned future cities project may bridge the gap between Japanese smart community development and an ideal one including the waste and transportation sectors.

6.3.2 *Project evaluation*

Regarding the evaluation of the projects, the goals of the four smart community projects were achieved overall. Table 6.6 displays the qualitative targets, the numerical targets, and the achievements of each smart community project. An evaluation of the results confirms that the smart community projects were successful as the achievements recorded in the four cities were greater than most of the numerical targets shown in the Table 6.6.

6.3.3 *User-driven innovation*

Using the final reports of the four smart community projects, it was revealed that societies were not involved in developing innovations. Table 6.7 shows all project participants in the four smart community

Table 6.6. The qualitative targets, numerical targets, and achievements of each smart community project.

(1) Yokohama

Qualitative targets	
Targets	**Achievements**
To develop an EMS for every 15 PJ of electricity which the project participants consume.	The goal was achieved.
To establish DR (demand response) programs in the building sector.	Maximum peak cut: 23%, and the DR (demand response) achievement rate was more than 90%.
To conduct a HEMS social trial in the whole area of Yokohama City.	The positive effects of dynamic pricing in electricity and HEMS promotion measures were validated.

Numerical targets		
Items	**Targets**	**Achievements**
Carbon dioxide emissions	30,000 t-CO_2	39,000 t-CO_2
Carbon dioxide reduction	25%	29%
Peak cut	20%	Maximum peak cut: 23% (for buildings) Net peak cut: 15.7% (for households)
Energy saving	17%	17%
PV installation	27MW	37MW
HEMSs installed in houses	4,000	4,230
EV (newly purchased)	2,000	2,294

(2) Toyota

Qualitative targets	
Targets	**Achievements**
To optimize energy use and traffic demand and supply in the household sector in consideration of consumer behavior.	The positive effects of HEMSs and other smart equipment, next-generation environmental automobiles, and new social technologies on infrastructure were all verified.

Table 6.6. (*Continued*)

Qualitative targets	
Targets	**Achievements**
Models developed for the low-carbon society system should be transferred to domestic markets and overseas.	The Ecoful Town built during the smart community project highly attracted domestic and foreign visitors. The number of visitors was approximately 180,000 from more than 100 countries.

Numerical targets		
Items	**Targets**	**Achievements**
Carbon dioxide reduction	30% (for the household and transport sectors)	30.1%

(3) Keihanna

Qualitative targets	
Targets	**Achievements**
In response to the 2011 Tohoku Earthquake and Tsunami, to identify how to cope with demand response (DR) (namely, peak cut) in times of a limited supply of electricity.	The combination of the local battery with the shortened DR times had a significant effect on the peak cut.
In times of sharply increasing volumes of renewable energies, especially PV, to identify how to manage surplus renewable energy.	Confirmed that EVs, which are not normally charged during the daytime, are a resource for surpluses of renewable energy.
To identify how to manage dynamic pricing by 30 minutes?	The combination of DR and battery storage was effective for dynamic pricing.

(*Continued*)

Table 6.6. *(Continued)*

Numerical targets		
Items	**Targets**	**Achievements**
Carbon dioxide reduction	35%	35%
Peak cut (1 pm to 4 pm, Summer)	28%	39%
Peak cut (6 pm to 9 pm, Winter)	42%	45%

(4) Kitakyushu

Qualitative targets	
Targets	**Achievements**
Citizen participation Through the installation of energy efficient equipment and energy-saving activities, to change regular consumers into production consumers (prosumers) that contribute to the supply of regional energy.	It was confirmed that members of the society joined the regional energy management, and the eco-tours and information disclosure enabled citizens to raise their awareness of energy.
Hardware installation To construct a new energy system, namely community energy management systems (CEMSs), for such "thinking" and "participating" prosumers.	Data gathered and analyzed through CEMS and smart meters were distributed to customers.
Social system and institution To establish a mechanism providing both residents' benefits and regional energy systems in terms of visualization of local energy information and dynamic pricing.	The demand response system based on past consumer electricity prices was examined, and an appropriate pricing system was set up. Statistically significant institutional analyses were performed using the CEMS data.

Table 6.6. (*Continued*)

Numerical targets		
Items	**Targets**	**Achievements**
Carbon dioxide reduction	50%	51.50%
Energy saving	20%	40.50%
Peak cut	15%	102.60%

Source: Ministry of Economy, Trade, and Industry [2016].

Table 6.7. Project participants in each smart community project.

(a) Yokohama Smart City Project (YSCP)	
Governments	City of Yokohama, Urban Renaissance Agency
Academia	Tokyo Institute of Technology
Industry	MM42 KaihatsuTMK (a subsidiary of Marubeni Corporation), Yokohama Smart Community, Accenture, NTT DoCoMo, NTT Facilities Inc., ORIX Corporation, ORIX Auto Corporation, Sharp Corporation, JX Nippon Oil & Energy Corporation, Sumitomo Electric Industries Ltd., Sekisui House Ltd., Sony Energy Devices Corporation, Daikyo Astage Inc., Taisei Corporation, Tokyo Gas Co. Ltd., TEPCO, Toshiba Corporation, Nissan Motor Co. Ltd., JGC Corporation, JGC Information Systems Company Ltd., NEC Corp., Nomura Real Estate Development Co. Ltd., Panasonic Corporation, Hitachi, Ltd., Misawa Homes Co. Ltd., Mitsui Fudosan Co. Ltd., Mitsui Fudosan Residential Co. Ltd., Mitsubishi Estate Co. Ltd., Meidensha Corporation
Citizens	4,230 households
(b) Toyota City Low-Carbon Society Verification Project (Smart Merit)	
Governments	Toyota City, Aichi Prefecture
Academia	Nagoya University

(*Continued*)

Table 6.7. (*Continued*)

Industry	Aishin Seiki Co. Ltd., Eneres Co. Ltd., KDDI Corporation, Circle K Sunkus Co. Ltd., Sharp Corporation, Shinmei Engineering Consultants Co. Ltd., Chubu Electric Power Co. Inc., DENSO Corporation, Toshiba Corporation, Toho Gas Co. Ltd., Toyota Motor Corporation, Toyota Industries Corporation, Toyota Chamber of Commerce and Industry, Toyota Smile Life Inc., Toyota Tsusho Corporation, Toyota Home, Dream Incubator Inc., Central Nippon Expressway Company Ltd., Nagoya Railroad Co. Ltd., Development Bank of Japan Inc., Hewlett-Packard Japan Ltd., Hitachi Ltd., Fujitsu Ltd., Mitsubishi Corporation, YAZAKI Corporation, Yamato Transport Co. Ltd., Yamaha Motor Co. Ltd.
Citizens	67 households
(c) Keihanna Eco City Next-Generation Energy and Social System	
Governments	Kyoto Prefecture, Kizugawa City, Kyotanabe City, Seika Town, Urban Renaissance Agency, (Kansai Research Institute), (Doshisha Yamate Sustainable Urban City Council)
Academia	(Kansai Research Institute), (Doshisha Yamate Sustainable Urban City Council)
Industry	(Kansai Research Institute), (Doshisha Yamate Sustainable Urban City Council), Kansai Economic Federation, Kyoto Center for Climate Actions, Enegates Co. Ltd., i-Energy WG, Osaka Gas Co. Ltd., Renesas Electronics Corporation, OMRON Corporation, The Kansai Electric Power Company Incorporated, Keihanna Interaction Plaza Inc., Sharp Corporation, Nihon Unisys Ltd., Mitsubishi Motors Corporation, Mitsubishi Heavy Industries, Ltd., Mitsubishi Corporation, Mitsubishi Electric Corporation, Fuji Electronic Co. Ltd., Furukawa Electric Co. Ltd., The Furukawa Battery Co. Ltd., Renesas Electronics Corporation

Table 6.7. (*Continued*)

Citizens	Approximately 700 households
(d) Kitakyushu Smart Community Project	
Governments	Kitakyushu City, (Human Media Creation Center/Kyushu)
Academia	(Human Media Creation Center/Kyushu)
Industry	(Human Media Creation Center/Kyushu), Azbil Corporation, lwatani Corporation, Uchida Yoko Co. Ltd., ORIX Corporation, Saibugas Co. Ltd., JXN Engineering Co. Ltd., NS Solutions Corporation, Nippon Steel & Sumitomo Metal Corporation, Sekisui Chemical Co. Ltd., Softbank Telecom Corp., Daiwa House Industry Co. Ltd., DENSO Corporation, TOTO, Toppan Printing Co. Ltd., Toyota Motor Corporation, Toyota Industries Corporation, Toyota Tsusho Corporation, Toyoda Gosei Co. Ltd., Nippon Steel & Sumikin Texeng. Co. Ltd., IBM Japan, Japan Telecom Information Service Co. Ltd., Higashida-Clinic, FamilyMart Co. Ltd., Fuji Electric Co. Ltd., Furukawa Electric Co. Ltd., The Furukawa Battery Co. Ltd., Hohkohsya, Mitsubishi Heavy Industries Ltd., Yaskawa Electric Corporation, Yaskawa Information Systems Corporation
Citizens	Approximately 180 households

Source: Japan Smart Community Alliance [2015].

projects. This table includes the municipalities and prefectures where the smart community projects were implemented and several organizations involved in the government–academia–industry collaboration. While many private companies participated in the smart community projects, no university in either Keihanna City or Kitakyushu City was involved with the projects as academia. Regarding citizens' participation, PV panels, EVs, and HEMSs were newly installed in general houses and business establishments in the four project areas. It seems clear, then, that society played a part in the smart community projects. It is noteworthy that citizens and business establishments participated in the smart community project as production consumers (prosumers) in Kitakyushu City. However, it should be emphasized that citizens' participation in the four smart community projects was limited to the installation process of existing innovation as users, and did not extend to the creation process of new innovation as collaborators.

6.4 Towards User-driven Innovation in Smart City Initiatives

Using the Quadruple Helix Model, this chapter aims to examine user-driven innovation in the smart community projects implemented in Japan between 2010 and 2014. Although all of the projects were evaluated as being successful overall, citizens' participation was not observed in the creation process of new innovation; therefore, user-driven innovation seemed to be limited regarding smart city development in Japan. Borkowska and Osborne [2018] examined the validity of the Quadruple Helix Model for the smart city project implemented under the Future City Demonstrator Initiative in Glasgow, Scotland. They revealed that various ICTs, such as smartphone apps and digital platforms, were introduced in this initiative; however, these technologies were not prepared to support citizen participation. Besides, a lack of digital literacy in a large part of Glasgow City led to citizens' having problems with the aforementioned technologies, and, consequently, the learning process among the four helixes, which is essential for innovation [Brown and Duguid, 1991; Powell *et al.*, 1996] was also hindered in this initiative. Looking closely at the four smart community projects in Japan from the viewpoint of supportive ICTs for citizen participation, no ICTs were introduced to promote citizen participation. In these projects, citizens were regarded as users or prosumers in the energy management system solution proposed by private companies. Accordingly, citizens were not given a proactive role. For these reasons, it seems that neither learning processes were created, nor was the participation of citizens as co-creators observed in these projects. Thus, in future, when developing a smart city based on the conceptual framework of the Quadruple Helix Model, the following two things may be necessary—(1) technological support, namely ICTs encouraging citizens' participation and (2) organizational assistance such as living labs that foster the learning process.

References

Albino, V., Berardi, U., and Dangelico, R.M. (2015). Smart cities: Definitions, dimensions, performance, and initiatives, *Journal of Urban Technology*, *22*, 1–19.

Bibri, S.E., and Krogstie, J. (2017). Smart sustainable cities of the future: An extensive interdisciplinary literature review, *Sustainable Cities and Society*, *31*, 183–212.

Borkowska, K., and Osborne, M. (2018). Locating the fourth helix: Rethinking the role of civil society in developing smart learning cities, *International Review of Education*, *64*, 355–372.

Brown, J.S., and Duguid, P. (1991). Organizational learning and communities-of-practice: Toward a unified view of working, learning, and innovation, *Organization Science*, *2*, 40–57.

Caragliu, A., del Bo, C., and Nijkamp, P. (2009). Smart cities in Europe, *Proceedings of the 3rd Central European Conference in Regional Science*, 45–59.

Carayannis, E., and Campbell, D. (2009). 'Mode 3' and 'Quadruple Helix': Toward a 21st century fractal innovation ecosystem, *International Journal of Technology Management*, *46*, 201–234.

Gibson, D.V., Kozmetsky, G., and Smilor, R.W. (eds.). (1992). *The Technopolis Phenomenon: Smart Cities, Fast Systems, Global Networks*. New York: Rowman & Littlefield.

Japan Smart Community Alliance. (2015). *SMART COMMUNITY—Japan's Experience.* Tokyo: Japan Smart Community Alliance.

Jorna, F., and Veenstra, M. (2015). Setting-up smart cities ecosystems—Essential building blocks, in: Kommer, P., and Peng, G.C. (eds.), *Proceedings of the International Conferences on ICT*, Society and Human Being 2015, Web Based Communities and Social Media 2015 and Connected Smart Cities 2015, pp. 140–146. Lisbon: IADIS Press.

Komiyama, Y. (2012). *Smart Communities, as a Presentation of US–Japan Renewable Energy Policy Business Roundtable "New Opportunities for Renewable Energy in Japan & the United States"*. URL: http://www.enecho.meti.go.jp/notice/event/036/pdf/event036013.pdf. Accessed: September 28, 2018.

Komninos, N. (2011). Intelligent cities: Variable geometries of spatial intelligence, *Intelligent Buildings International*, *3*, 172–188.

Komninos, N., Pallot, M., and Schaffers, H. (2013). Special issue on smart cities and the future internet in Europe, *Journal of Knowledge Economy*, *4*, 119–134.

Lazaroiu, G.C., and Roscia, M. (2012). Definition methodology for the smart cities model, *Energy*, *47*, 326–332.

Lea, R., Blackstock, M., Giang, N., and Vogt, D. (2015). Smart cities: Engaging users and developers to foster innovation ecosystems, *Proceedings of the 2015 ACM International Joint Conference on Pervasive and Ubiquitous Computing and the Proceedings of the 2015 ACM International Symposium on Wearable Computers,* September 7, 2015, pp. 1,535–1,542. New York: ACM.

Martinez, S., Bjerkås, S., Ludvigsen, A.E., and Fensli, R. (2016). Agder living lab: From ideas to large-scale deployment and long term user adoption of inclusive health solutions, *Advances in Intelligent Systems and Computing*, *500*, 391–399.

Ministry of Economy, Trade, and Industry. (2011). *Documents of the 13th meeting of the Next Generation Energy and Social System Council*. URL: http://www.meti.go.jp/committee/summary/0004633/013_haifu.html. Accessed: September 28, 2018.

Ministry of Economy, Trade, and Industry. (2016). *Documents of the 18th meeting of the Next Generation Energy and Social System Council*. URL: http://www.meti.go.jp/committee/summary/0004633/018_haifu.html. Accessed: September 28, 2018.

Ministry of Economy, Trade, and Industry. (2017). *The Concept of a Smart Community*. URL: http://www.meti.go.jp/english/policy/energy_environment/smart_community/index.html. Accessed: September 28, 2018.

Mora, L., Deakin, M., and Reid, A. (2018). Strategic principles for smart city development: A multiple case study analysis of European best practices, Technological Forecasting and Social Change (in press).

OECD. (2013). *Green Growth in Kitakyushu*, Japan, Paris: OECD.

Powell, W.W., Koput, K.W., and Smith-Doerr, L. (1996). Interorganizational collaboration and the locus of innovation: Networks of learning in biotechnology, *Administrative Science Quarterly*, *41*, 116–145.

Schaffers, H., Komninos, N., Pallot, M., Trousse, B., Nilsson, M., and Oliveira, A. (2011a). Smart cities and the future internet: Towards cooperation frameworks for open innovation, in: Domingue, J., *et al.* (eds.), *The Future Internet*, (FIA 2011, Lecture Notes in Computer Science, vol. 6656), pp. 431–446. Berlin, Heidelberg: Springer.

Schaffers, H., Sällström, A., Pallot, M., Hernández-Muñoz, J.M., Santoro, R., and Trousse, B. (2011b). Integrating living labs with future internet experimental platforms for co-creating services within smart cities, *Proceedings of 2011 17th International Conference on Concurrent Enterprising*, 1–10.

Shigematsu, C. (2015). *Creation of Smart Communities in Keihanna Science City—From Demonstration to Implementation*. URL: https://www.smart-japan.org/english/vcms_cf/files/Keihanna_Project_English.pdf. Accessed: September 28, 2018.

Tobe, C. (2015). *Japan's Policy on Smart Communities*. URL: https://sustainabledevelopment.un.org/content/documents/12634tobe.pdf. Accessed: September 28, 2018.

© 2023 World Scientific Publishing Company
https://doi.org/10.1142/9789811257186_0007

Chapter 7

Science and Technology Park Development in the Context of Smart Cities

Raimund Bröchler,[*,‡] Mathias Seifert[†]

[*]*INTRASOFT International S.A., Luxembourg, and INSME, Rome, Italy*
[†]*Hochschule Fresenius, University of Applied Science, Berlin, Germany*
[‡]*Corresponding author, Email: raimund.broechler@intrasoft-intl.com*

7.1 Introduction: Smart Cities

Cities have always faced challenges and opportunities, but nowadays, cities are urban spaces which face a never-before-seen complexity of challenges, but also very promising opportunities [Acuto and Steele, 2013]. To create the solutions, we need to go beyond known city governance models and subsystem architectures, and break down a lot of siloes. The transformation requires new governance models with changing roles and responsibilities, a much higher level of interoperability not only on the information technology (IT) level but also between all those siloed solutions and services that have been created before. Our focus in this chapter will be mainly to pinpoint some crucial interoperability challenges observed, and they are not 'just' technical issues; a big part of these changes has to be addressed on a policy level, technical level, service level, stakeholder engagement level, authority level and funding level. Science Parks (SP) or Science and Technology Parks (STP), for example, are still an appreciated policy instrument created on the regional state level or state level and therefore specific 'city-level' needs, 'customization' of services requests, or provision of city-level testbeds cannot be addressed, as city governance has limited administrative power in these projects and sometimes they are not even involved as stakeholders of the STPs. Connected to this misfit, the technology orientation of the STPs and engagement in topics that are of core interest of the Smart Sustainable City (SSC) are not fitting at all to the orientation of how the STP is connected to the city. According to the MGI Smart Cities report [McKinsey Global Institute, 2018], such key domains that affect multiple aspects of the quality of life and sustainability include mobility, security, healthcare, energy, water, waste, economic development and housing, as well as engagement and community.

7.2 Science and Technology Parks and Smart Sustainable City Governance: Is There a Need to Better Connect So-Far Siloed Solutions?

Amazingly enough, the literature offers many new articles and studies covering SPs or STPs as well as a lot of literature about SSCs. We see SCCs as a necessary evolution of the Smart City (SC) concept, putting more emphasis on the citizens and aspects of quality of life than purely on technology and resource efficiency. The same is true for SPs and STPs; STPs are also an evolution of the former purely SP concept, taking open innovation and exploitation tasks into account; but not all authors share this view, and so we will always talk about SPs/STPs. Despite the fact that there are many literature available, there are not many articles where the typical tasks and goals of SPs/STPs are described as a core ingredient in the new SSC concept, except Carvalho [2017], who analyses the entrepreneurial development and innovation within the SC, and Zouain and Plonski [2015] who ask what makes STPs contribute to improving the quality of urban development, analyzing STPs and their role in Brazil as a case study.

This chapter would like to lobby for an approach to have STP governance much more localized and being under or a part of SSC governance. The technological competence can still be wider than the SSC demand but must at least cover the SSC demands fully. The funding can still come from the regional state or national level, but in order to be much more interconnected regionally and more efficient, SSCs have to be the key player and client of technological input coming from STPs. This is also important in order to create more innovation besides the offers of the large corporations that already dominate this market with big vendor log-in risk of their solutions. To be clear, we do not want to replace Ministry-level program ownership and governance to the Major level as we believe the private sector also plays a key role in proper SCC governance models like citizens, while Ministries and Majors act as enablers. Technology is only as effective as the entity that puts it to work [Genta and Riberi, 2019].

A split in public policies will occur because public authorities must be capable of "thinking city" and developing new abilities in system architecture. This implies a profound evolution of the administrative organization, which must operate transversally, as it is impossible to think in terms of complex systems with an administration compartmentalized into silos [Rochet, 2018]. A very good example and maybe benchmark is Singapore. Singapore found an articulation between the central role of the government and the initiative of actors; a city is an archetype of the strategist state. In Singapore, the atmosphere is favorable to pilot and innovative field projects that are quickly integrated into the system of systems [Rochet, 2017]. The exercise to be done in all those STPs and SSCs is to take the existing subsystem architectures and concepts and make them in a way interoperable as a system of systems, that they function properly together, reusing as much as possible from what exists (measured according to key performance indicators, or KPIs) from all those services, functions and features in order to capitalize on the already made investments, while carefully breaking all siloes that hinder interaction between the stakeholders and business models. We believe STPs (or at least the aim of any SC/STP to foster entrepreneurial development and innovation) must be better integrated in the Smart Sustainable City concept and business model, as STPs are not merely investments. They can heavily increase the impacts of innovation when fully integrated within the SCC concept by taking over the key features of the innovation front-end. A best fit of such an approach can be seen in the Barcelona Urban Lab (BUL),[1] which is a tool to facilitate the use of public spaces in the city of Barcelona to carry out tests and pilot programs on products and services with an urban impact. This is in the pre-market stage and is in line with the Barcelona City Council's aims, priorities and lines of action. Through the BUL project, the city is made available to companies with innovative projects for the future to test their infrastructures and services in a real environment. This concept fits very well in the European Program of Digital Innovation Hubs (DIH).[2] DIHs are one-stop shops that help companies to become more competitive with regards to their business or production processes, products or services using digital technologies. Being an evolution of the Smart Specialization activities in Europe, the DIH concept is a very good approach for how STPs could reinvent themselves and fine-tune themselves for being SSC-interoperable.

What we especially need for developing countries is a smart, lean, agile and sustainable concept that is no longer copying the 1980s phenomenon of the North American Research Park model, and follow only the newest models that are properly adapted to local contexts. Those former mostly greenfield models[3] do not fit anymore in modern sustainability concepts; STPs must be integrated service providers within SSC concepts, being close to the cities and avoiding wholesale access by private cars. There is a clear trend in all healthy startup-friendly cities where incubators move from the periphery to the areas that are close to or into the vibrant city life and cultural activities; not only to be more competitive and attract global talent. Incubators often use old industrial or harbor areas to offer their services. These areas are thus becoming very attractive parts of the changing cities. Attracting startups and small- and medium-sized enterprises (SMEs)

---

[1] URL: https://www.centreforpublicimpact.org/case-study/barcelona-urban-lab.

[2] European Commission. Digital Innovation Hubs. URL: https://s3platform.jrc.ec.europa.eu/digital-innovation-hubs.

[3] Greenfield STP models or projects are new projects which are built from scratch on greenfield land while Brownfield projects are those facilities which are based on modified/upgraded existing projects.

within STPs which are greenfield located is becoming more challenging, because of its very low service density, precludes the emergence of support facilities, and further reliance on the private car is engendered, particularly during the lunch hour [Why, 2001]. Having successful STP and SSC implementation is an integration problem, as it breaks silos of responsibilities and asks for interoperability of partially already existing IT and Internet of Things (IoT) infrastructure. Only a governance model that goes beyond an existing group of stakeholders can manage this. Rochet [2018] defines a smart city as a 'system of systems' and proposes a concept with an integrative line of thinking which avoids the pitfall of reducing a set to its sub-set—a city is not just its economy, its culture, traffic, energy, housing, etc., but it is the integration of all of these elements. To the entrepreneur, it offers a perspective on their activity that likens it to a rock in an edifice which is bigger than him or her, a great creation that inspires and draws them in and gives them meaning. To the citizen, it offers a path to reconstruct the link between the common good and an individual one, which was the basis for the prosperity of medieval towns and cities [Rochet, 2018]. Parry [2018], also referring to the Brookings Institution [Katz and Wagner, 2014], describes several of those components like Economic Asset, Physical Asset and Network Asset, that must be connected for a fully functioning SP ecosystem, and many of those provisions fall under urban governance. All those successful cities with hundreds of years of experience have already faced those evolutions many times and this is how we should tackle the problem as well. It is very close to the problem that the manufacturing industry is facing when reaching Industry 4.0. The pure new technology aspects are still important, but only as one of many aspects that have to be taken into account. Nowadays, the key aspect is interoperability between technology solutions and services, but also within the governance layers, where we still see too many islands and siloed solution approaches. This approach addressing the key issue of how to break down silos of services and functions to create real Smart Cities was tackled by several reference cities like Stockholm or Lisbon (Ferreira, 2021).

7.3 Science and Technology Parks and Smart Sustainable City Concepts and Development: How Can They Benefit from Each Other?

STPs have become a very popular policy tool to enhance knowledge-based regional development since the early 1980s. STPs can be found all over the globe based on different models of stakeholder involvement and mission [Geenhuizen and Soetanto, 2008]. The first Science Parks emerged in the US in the 1950s at Stanford Research Park and Research Triangle Park, and the first Science Park in Europe was established in Edinburgh in 1965. In 1970, there were only 21 Science Parks throughout the world [Haxton, 1998]. This number increased to 473 science and research parks in 1998 [McQueen and Haxton, 1998], and in 2014, Europe counted about 400 Science Parks [Rowe, 2014]. This amazing increase of Science Parks and STPs can only be explained by a kind of common policy understanding where STPs are the Number 1 tool used to encourage the formation of knowledge-based businesses and other organizations that are normally residents on-site [Ylinenpää, 2001].

Despite the fact that such growth rates where Science Parks and STPs have become common practice, its impacts are still far from being understood and are often debated [Lecluyse *et al.*, 2019]. Geenhuizen and Soetanto [2008] even state: "Despite the rather poor proof of success, Science Parks, paradoxically, have remained extremely popular as a policy tool". In times of open innovation, full digitalization of communication from telcos, meetings to social media, digitalization of industry and globalization, are SPs with localized tenants still the right instrument? These days, does 'close' refer to distance or does it refer to the frequency of interaction that does not care about distance? However, in the event that being close to the end clients is an important element, then we need to fully capitalize on this and use the testbed functionality much more for Smart Sustainable City solutions. Did SPs and STPs miss digitalizing their services, missed the STP 4.0 evolution, and did they miss detecting a core client of their services and functionalities, the SSCs? Is this one of the reasons why we get a very diversified picture of the performance when analyzing SPs or STPs? As reflected by Nararasimhalu [2013], STPs are key partners within the Open Innovation Ecosystem and we suggest adding SSCs as a key partner in the presented valorization model. Valorisation of high-quality infrastructure and services of many intermediaries does not properly function, as we often miss the high-level orchestration of the national, regional and city-level policy development, and therefore we see missing interoperability within the support services. Innovation infrastructures and digitalization services are boxed in silos and thus miss out on the interoperability of infrastructure, services as well as on technical level, whereas its impact could be increased with better coordination and a meta-level governance of those instruments.

SPs or STPs are a part of policy tools that serve to enhance the knowledge-based economic growth of regions. Important policy tools that are used in combination are intermediary organizations for knowledge and technology transfer, networking and cluster programs, and finally investment support schemes. Since the 1980s, and depending very much on the stakeholders and regional context, there are attributes that all SPs and STPs have in common [Gower and Harris, 1994]:

- a property-based initiative close to a place of learning (university or research institute);
- the supply of high-quality premises or units to businesses; and
- a policy context of mixed public/private stakeholders with expectations on the knowledge-based economic results of the parks.

In order to find commonalities between SPs or STPs and Smart and Sustainable Cities, we should compare the contributors for each of them. According to Lecluyse *et al.* [2019] the contributors could be categorized in Inputs (regional level, SP/STP-level, firm-level), Mediators (networking for SP/STP and firm-level) and Outputs (regional level outcomes, firm-level outcomes).

SCs or SSCs both do not yet have a normalized definition. So, we mainly have prototypes that we take in order to define them. Some of them are showcases for technologies endorsed by our largest tech companies (Cisco, IBM, Siemens, Microsoft, etc.), such as Songdo (Korea), Masdar (Abu Dhabi) and Plan IT Valley (Portugal), but these are not cities made to be lived in by real people [Rochet, 2018]. After a decade of trial and error, municipal leaders are realizing that smart city strategies start with people, not technology. "Smartness" is not just installing digital interfaces in traditional infrastructure or streamlining city operations. It is about using technology and data purposefully to make better decisions and deliver a better quality of life [McKinsey Global Institute, 2018]. However, looking at the expansion of Singapore [Rochet, 2017], we can see that absorbing hard technologies is actually relatively easy, and as soon as a country has understood that the keys to success are in the intellectual investment of soft technologies and social capital, the student overtakes the master in only a matter of decades.

As previously mentioned, the key aspect is to master the hard technologies and the soft technologies. Whoever masters the ecosystem of the city will de facto master the underlying technologies, and will define the demand, while the supply also defines standards [Rochet, 2018]. This is a promising role to have, not only because the SC/SSC market is

Figure 7.1. The chinese approach to urban integration.
Source: Quangbin Wang, in: Rochet [2018]; see also for a larger version of this figure: www.iste.co.uk/rochet/cities.zip.

forecasted in 2020 with having a valuation in studies of $1.5 billion with 70% in infrastructure and about 30% in services. Chinese planners have not wasted any time integrating such a system approach, as visualized in Figure 7.1.

The advantage of this Chinese strategy is in looking at a city as a whole (i.e., a system of systems approach) rather than a mere sum of its parts, accentuating the ability to integrate technologies which, for the time being, come from the Western world. Their approach understands that what controls the system controls the underlying technologies, and not the other way around.

7.4 Smart Sustainable Cities Supporting Entrepreneurship and Innovation, the Integration Challenge

As previously mentioned, the biggest cities worldwide will be the most important testbeds for new sustainability approaches, but also testbeds for better transactions of communication, processes, business models, inter-governmental collaboration and the destruction of former siloed approaches with very low interoperability, and this is meant technically, process- and communication-wise. Most of these cities have hundreds of years' experience in reinventing themselves and already saw three industrial revolutions. The next challenge is the secured full digitalization, combined with a much more participative collaboration with the citizens.

Those changes are disruptive changes, changes that are not done as greenfield approaches on a drawing table. The changes must be implemented within a running and existing system like a customization approach because of its new requirements. The process is for sure a loop process which is, in a way, never-ending permanent learning—the Möbius strip. Integrating STPs fully into SSC concepts mean an extension of the already known value propositions with new stakeholders and new governance models that are different to what is practiced so far. Comparing this with Parry's governance models [Parry, 2018], the approach could follow a Startup/Entrepreneur-led cluster approach

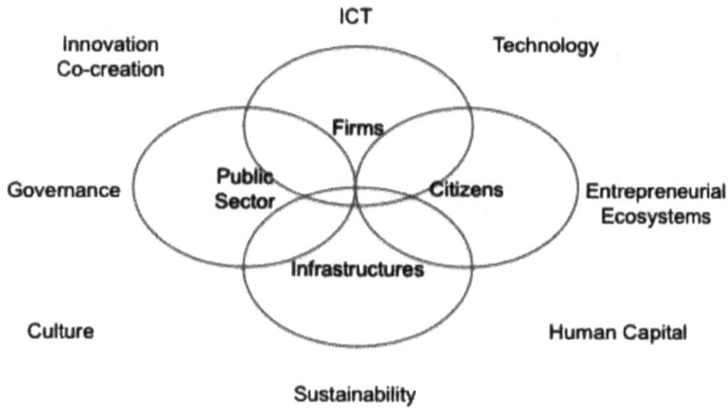

Figure 7.2. Smart city environment.
Source: Carvalho [2017].

that is also described in SSC methodologies, where the integration is Public–Private Partnership-driven, and Governments are just preparing and facilitating the environment for proper integration. Carvalho [2017] defines a smart city according to the mandatory environment where the typical core elements of SPs or STPs are integrated (Figure 7.2).

Normally, an SP or STP contributes to three (out of eight) areas shown in Figure 7.2, Innovation and Co-creation, Technology and Entrepreneurial Ecosystems. As most of the Cities already exist and are not set up on a greenfield, the right model and strategy have to be selected, depending on the objectives, motivation, governance model, partners, funding, etc. This means that SCs/SSCs can be planned and projected or originated from retrofitting projects. Of course, the actual maturity level (disperse, integrated or connected) is important as well and depends on the local and required specificities.

In our point of view, SPs or STPs must undergo a two-stage transformation in order to be ready to be part of a system of system approach. The first stage is transforming the SP or STP in a Digital Innovation Hub (DIH) ecosystem. The second stage is the full integration of the newly-created DIH into a holistic concept of SCs/SSCs, overcoming former bottlenecks and inconsistencies of siloed policy solutions that separated the concepts, solutions and services without any interoperability between them.

Digital Innovation Hubs as a first transformation step are one-stop shops that help companies become more competitive with regards to their business or production processes, products or services using digital technologies. They are based upon technology infrastructure (Competence Centre, or CC) and provide access to the latest knowledge, expertise and technology to support their customers with piloting, testing and experimenting with digital innovations. DIHs also provide business and financing support to implement these innovations, if needed, across the value chain. As proximity is considered crucial, they act as a first regional point of contact, a doorway, and strengthen the innovation ecosystem. A DIH is a regional multi-partner cooperation (including organizations like Research and Technology Organisations (RTOs), universities, industry associations, chambers of commerce, incubator/accelerators, regional development agencies and even governments), and can also have strong linkages with service providers outside of their region supporting companies with access to their services. The model is visualized in Figure 7.3. This approach also takes into account the former challenge mentioned that SPs or STPs are normally policy tools that are created on a federal-state level or even country-level in small countries.

DIHs create with their embedded design and architecture an entrepreneurial ecosystem, and this links also to the former mentioned second integration step. Entrepreneurial Ecosystem is a recent concept without a clear consensus about its definition. Isenberg [2010] proposed nine 'prescriptions' for creating an entrepreneurial ecosystem:

(1) *Stop Emulating Silicon Valley*: While Silicon Valley is a successful entrepreneurial ecosystem, it is unique for its region and unlikely to be replicated in other areas.
(2) *Shape the Ecosystem around Local Conditions*: Look for locally-based industries with growth potential and existing capacity, and build upon these foundations.

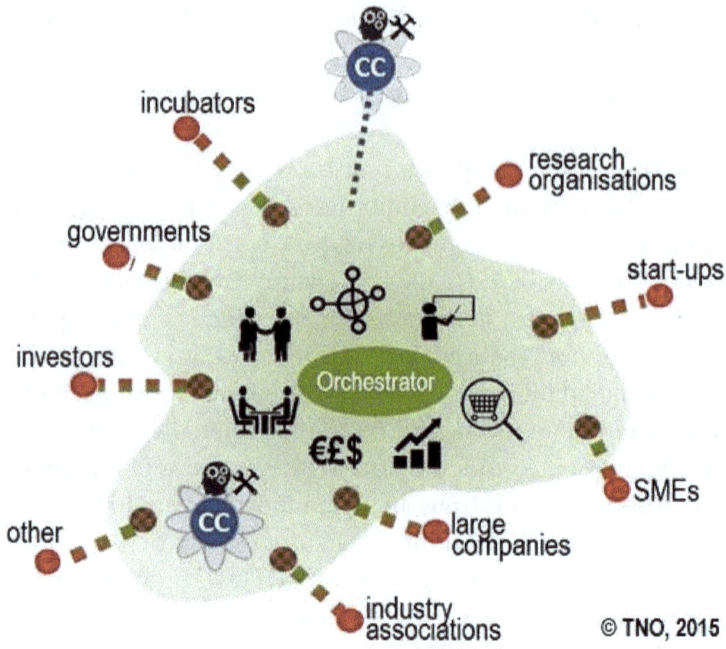

Figure 7.3. Digital innovation hub ecosystem.
Source: Roundtable on Digitising European Industry: Working Group 1–Digital Innovation Hubs[4] [2017].

(3) *Engage the Private Sector from the Start*: Entrepreneurial ecosystems must be led by the private sector and the role of the government is to facilitate, not to try to lead or control.
(4) *Favor the 'High Potentials'*: While there must be room for all types of businesses, attention should be given to fostering the growth of firms with the capacity for innovation and global market engagement.
(5) *Get a Big Win on Board*: Success stimulates and motivates others to have a go, and where there are successful firms, they should be showcased and used as case examples for others.

---

[4] European Commission. (2017). *Roundtable on Digitising European Industry*. URL: https://ec.europa.eu/futurium/en/system/files/ged/dei_working_group1_report_june2017_0.pdf.

(6) *Tackle Challenges Head-On*: Entrepreneurial activity in some areas may be stifled by an entrenched culture that is risk-averse or conservative; this should be challenged by active communication and education programs.
(7) *Stress the Roots*: Entrepreneurial growth-oriented firms should not be flooded with 'easy money' through grants or venture capital flows. Firms must be profitable and sustainable with good financial management.
(8) *Do not over Engineer Clusters; Help Them Grow Organically*: Government enthusiasm for building industry clusters needs to be tempered by a realization that they emerge organically from existing industries and not from attempts to 'pick winners' or build greenfield science parks.
(9) *Reform Legal, Bureaucratic, and Regulatory Frameworks*: A key role for the government is to address legal, bureaucratic and regulatory issues such as taxation, licensing and compliance, so that there are no unnecessary impediments to entrepreneurship and small business growth.

SPs or STPs is a policy instrument and very often they do not respect those prescriptions. In those government-driven approaches, SPs or STPs are often built on promising new technology areas without having the locally-based industries with growth potential and existing capacity to build upon these foundations. Large exogenous firms also play a major role in developing regional ecosystems, especially in peripheral regions, developing the ecosystem's managerial talent pool and providing commercial opportunities for local businesses. Additionally, large companies can also give several other contributions, including the provision of space and resources for local startups, the creation of programs to support startups and the development of companies that improve their own ecosystems.

7.5 Conclusion

This chapter lobbies for the revision of the traditional features and services of Science Parks (SPs) and Science and Technology Parks (STPs).

SPs and STPs have optimized their features and services up to the limit the governance model has given them. However, this is exactly the problem; the limited interoperability to connect as a sub-system to the system of systems. On the other side, we have the Smart Cities (SCs) and Smart and Sustainable Cities (SSCs)—their governance system should allow for the orchestration of the system of systems and instead of reinventing the wheel, we should connect SPs or STPs as the innovation front-end; being part of the Innovation Ecosystem. We should be very careful to put our focus on a sustainable integration of optimized sub-systems. As the investments are massive, we should be very careful to integrate sustainable business models into this system of systems that are not only generating revenues just and only for large corporates and their solutions. SSCs are reinventing themselves every day according to new demands; a vendor lock-in system is dangerous because of missing flexibility and missing integration of business models and solutions provided by players that are part of the ecosystem. There are open-source solutions available that would support revenue creation for all players in the system like the FIWARE Smart City solutions (FIWARE Foundation, 2019). We need two-sided markets for SSCs and STPs transformed into modern DIHs that support the Innovation Ecosystem, as well as allow financial sustainability.

References

Acuto, M., and Steele, W. (2013). *Global City Challenges. Debating a Concept, Improving the Practice*. New York: Palgrave Macmillan.

Carvalho, L.C. (ed.). (2017). *Handbook of Research on Entrepreneurial Development and Innovation Within Smart Cities*, IGI Global, Hershey PA.

Ferreira, M.I.A. (2021). How smart is your city? *Technological Innovation, Ethics and Inclusiveness*. Cham: Springer.

FIWARE Foundation Berlin. (2019). *The Open Source Platform for Our Smart Digital Future*. Last modified December 2019. URL: https://www.fiware.org/community/smart-cities.

Geenhuizen, M.V., and Soetanto, D.P. (2008). Science parks: What they are and how they need to be evaluated, *International Journal of Foresight and Innovation Policy*, *4*(1–2), 90–111. URL: https://doi.org/10.1504/IJFIP.2008.016908.

Genta, G., and Riberi, P. (2019). *Technology and the Growth of Civilization*. Cham: Springer.

Gower, S.M., and Harris, F.C. (1994). The funding of, and investment in, British science parks: A review, *Journal of Property Finance*, 5(3), 7–18. URL: https://doi.org/10.1108/09588689410078557.

Haxton, B. (1998). Science Parks Around the World, *Facility Management Journal* March/April. Cited in: Ylinenpää, H. (2001). Science parks, clusters and regional development, *Paper presented at the 31st European Small Business Seminar in Dublin*, September 12–14, Luleå University of Technology.

Isenberg, D.J. (2010). How to start an entrepreneurial revolution, *Harvard Business Review*, *88*(6), 40–51. URL: https://institute.coop/sites/default/files/resources/Isenberg-HowtoStartanEntrepreneurialRevolution.pdf.

Katz, B., and Wagner, J. (2014). The rise of innovation districts: A new geography of innovation in America, *Metropolitan Policy Program at Brookings (May 2014)*. Washington: Brookings Institution. URL: https://www.brookings.edu/essay/rise-of-innovation-districts.

Lecluyse, L., Knockaert, M., and Spithoven, A. (2019). The contribution of science parks: A literature review and future research agenda, *The Journal of Technology Transfer*, *44*(2), 559–595. URL: https://doi.org/10.1007/s10961-018-09712-x.

McKinsey Global Institute. (2018). *Smart Cities: Digital Solutions for a More Livable Future*. McKinsey & Company. URL: https://www.mckinsey.com/~/media/McKinsey/Industries/Public%20and%20Social%20Sector/Our%20Insights/Smart%20cities%20Digital%20solutions%20for%20a%20more%20livable%20future/MGI-Smart-Cities-Executive-summary.pdf.

McQueen, D., and Haxton, B. (1998). Comparison of Science Park planning, economic policy, and management techniques between Science Parks, *Paper Presented at XV IASP World Conference on Science & Technology Parks*, Perth, Australia, October 18 to 23, 1998. Cited in: Ylinenpää, H. (2001). Science Parks, clusters and regional development, *Paper Presented at 31st European Small Business Seminar in Dublin*, September 12–14. Luleå University of Technology.

Narasimhalu, A.D. (2013). CUGAR: A model for open innovation in science and technology parks, *World Technopolis Review*, *2*(1), 10–20. URL: https://doi.org/10.7165/wtr2013.2.1.10.

Parry, M. (2018). The future of science parks and areas of innovation: Science and technology parks shaping the future, *World Technopolis Review*, *7*(1), 44–58. URL: https://doi.org/10.7165/wtr18a0430.18.

Rochet, C. (2017). Singapour: La voie des villes intelligentes, *Constructif*, *46*, 22–25. URL: http://www.constructif.fr/articles/numeros/pdf/constructif-46.pdf.

Rochet, C. (2018). *Smart Cities—Reality or Fiction*. ISTE, London, Wiley, Hoboken.

Rowe, D.N.E. (2014). Setting up, managing and evaluating EU science and technology parks—An advice and guidance report on good practice, *European Commission, Directorate-General for Regional and Urban Policy*, Brussels. URL: https://ec.europa.eu/regional_policy/sources/docgener/studies/pdf/stp_report_en.pdf.

Why, P.H. (2001). Science and technology parks—are they relevant today? *Industry and Higher Education*, *15*(3), 219–221. URL: https://journals.sagepub.com/doi/pdf/10.5367/000000001101295696.

Ylinenpää, H. (2001). Science parks, clusters and regional development, *Paper presented at 31st European Small Business Seminar*, Dublin, September 2001. URL: https://pdfs.semanticscholar.org/e0fe/ac0bc011a6d827e0e8cba9fa1c0008564512.pdf.

Zouain, D.M., and Plonski, G.A., (2015). Science and technology parks: Laboratories of innovation for urban development—an approach from Brazil, *Triple Helix*, *2*, 7. URL: https://doi.org/10.1186/s40604-015-0018-1.

© 2023 World Scientific Publishing Company
https://doi.org/10.1142/9789811257186_0008

Chapter 8

Korean Technoparks: An Evolving Integrated Business Platform

Jaehoon Rhee

Gyeongbuk Technopark, Korean Technopark Promotional Association, Yeungnam University, Gyeongsan, Korea
Email: jrhee@gbtp.or.kr

8.1 The Korean Experience with Science and Technology Parks (STPs) Policy

Before introducing technoparks, the Korean models of the Science and Technology Park (STPs), as an integrated business platform, like most countries that have developed university–industry–government (UIG) relations and incorporated some independent bodies, Korean governments have also tried to implement various industrial policies for technology development and various research complex projects to upgrade their research and development (R&D) bases. While some were found very successful, however, at times of inauguration, many of them turned out to be time-consuming and under-performing, as the social and economic conditions had not been conducive.

8.1.1 *National industrial complex policy*

In fact, until the late-1990s, successive Korean governments implemented various industrial policies for technology development, and some of them, such as the Economic Development Policies Consistent with Economic Discrimination (ED) principle [Jwa, 2017], were found to be very successful, while others did not work out.

In the early-1960s, following the inauguration of Chung-Hee Park as President and the launch of the Five-Year Economic Development Plans, Korea recorded rapid growth, unprecedented in any other parts of the world. National development efforts, such as the switch from import substitution to the export-led growth and the "selective and concentrated" industrialization policy under the Heavy and Chemical Industry (HCI) drive, are just a few of the many distinct policies responsible for Korea's rapid growth and development. The National Industrial Complex policy (Figure 8.1) was especially successfully introduced to accommodate those companies specializing in HCI. Of course, strong government intervention and guidance underlined the policy initiatives of Korea under President Park. As a result, a number of small- and medium-sized enterprises (SMEs) established in the 1960s in the national industrial complexes grew into big corporations in the 1980s, and are now internationally competitive conglomerates like Samsung, Hyundai, and LG, etc.

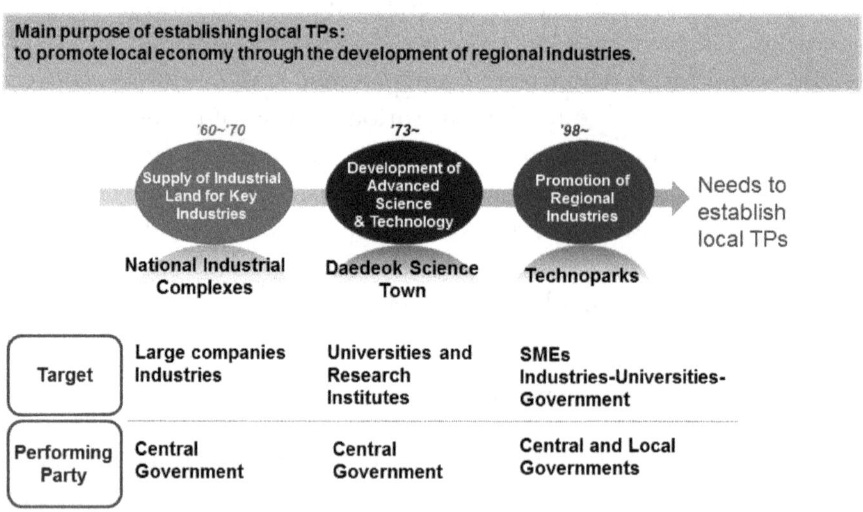

Figure 8.1. Korea's major national industrial complex policy.
Source: Excerpt from Rhee [2019a].

Of course, the growth of the Korean economy in the 1960s and 1970s has been attributed to various factors, but it is important nonetheless to note that the role of the government under the President Park's regime has been critical, acting as the main vehicle through which economically viable entities were selected and provided support, while those that were not deemed viable were removed from continued economic participation. The National Industrial Complex and Economic Discrimination policies were at the center of the export and economic growth strategy. With regard to specialized industrial complex policy, specifically the promotion of the HCI, the government maintained the same economically discriminatory policy mechanism by helping those firms that exhibited higher economic performances. Firms that demonstrated international competitiveness in their export activities by earning more dollars in the export promotion stage were allowed to enter the HCI sector with various government support programs, thereby increasing the success probability of the nations' industrial policy. The Korean industrial policy turned out to be a rare success case among 20th-century's history of industrial policy experiences in the world, thanks to this "government-led economically discriminatory industrial policy" [Jwa, 2017].

8.1.2 *Daedeok Science Town* (*or Park*) (*DST or DSP*)

However, President Park realized that Korea should have basic and applied R&D capabilities in order to catch up with developed countries' competitiveness, and improve and upgrade Korea's products. Therefore, President Park worked towards the historical formation of Daedeok Science Park (DSP) in the early 1970s.

The President Park government required the Ministry of Science and Technology (MST) to develop an industrial cluster in the Daejeon region, where social actors were easily able to get access to the nation's transportation network of highways and railroads. The MST had the responsibility to plan and coordinate the process of developing the STPs in Daejeon. and the Ministry of Education helped them to build new schools and relocated universities to that area. In the 1980s, some private research institutes of big conglomerates were also attracted to relocate to the park. Nowadays, a large number of research institutes within the complex of the DSP have developed and commercialized with many technologies, and cooperated with private sectors. As a result, the DSP has been positioned as a major axis of scientific research and development in Korea [Shin, 2001].

However, in actual fact, until the late-1990s, the DSP had not performed well despite enormous input from the government, only because of the low level of social capital like trust, a lack of strong infrastructure, and especially inexperience in connecting technology and commercialization.

8.1.3 *Technopark project*

Therefore, to overcome the limitations of the national industrial complex and the DSP's previous experiences, to replicate the success of Silicon Valley and to take the lead of both Taiwan and Singapore, which were particularly keen to invest in new STPs in an attempt to enhance economic competitiveness, the Korean government introduced the so-called Technopark project in 1998. The Korean government started the Technopark project in order to encourage and support startups and SMEs and promote regional industries. A technopark, the Korean version of a science and technology park, was originally designed as an under-one-roof supporting center for startups and SMEs by building infrastructures and providing equipment as well as an incubator. In particular, from the 1990s, the Korean government focused mainly on establishing high-tech industrial parks in order to develop SMEs and high-tech industries, prepare for a knowledge-based society and refresh the local economy. Further, the government implemented the business incubator program in order to revitalize local economies, create jobs and nurture high-tech industries.

It is also interesting to note that the then President Dae-Jung Kim government at the outbreak of the financial crisis in 1997 immediately put in place what were called the *chaebol* reform policies that demanded a corporate restructuring by way of increased management transparency, elimination of cross-debt guarantees, strengthening accountability to shareholders, reduced debt-equity ratios and specialization to "core" business lines under the IMF reform program as one of the conditions of the IMF rescue loan. Furthermore, the anti-*chaebol* reform policies were in fact strengthened in the midst of the crisis, when fingers were pointed at the *chaebol* for causing the crisis with over-borrowing and by letting the second-largest *chaebol* as well as the best exporting corporation, Daewoo Corp, go bankrupt in 1999 under the tight export credit crunch led partly by the government. More than a dozen mid-sized *chaebols* including the old Kia Motors, Hanbo Steel, etc., were dissolved in the crisis via the restructuring process.

In fact, it is hard in principle to blame the *chaebols* for the crisis, but the then Korean government and Korean people realized that *chaebols* could not help Korea get out of the crisis and could even make Korea go broke—that is, the Korean government and people were disillusioned with the so-called "too-big-to-fail" myth. In hindsight, the *chaebol* reforms during the crisis signaled the end of Chung-Hee Park's growth paradigm, or government-led economic discrimination, thereby destroying, to a large extent, the incentive for the growth of the corporate sector.

Since the financial crisis, Korea has experienced worsening growth and distribution following the short-lived rebound immediately after the crisis. With this background, entering the mid-2000s, social policy concerns began to dominate public debate, while growth and development policy issues have increasingly lost sight of economic discrimination, which has been overwhelmed by strengthening egalitarian ethos and policy.

This trend has been reinforced by most of the latter governments, the left-wing Moo-hyun Roh government (2003–2007) and the current left-wing Jae-In Moon government (2017–present). Regardless of their ideological difference, the general economic policy direction of Korea is now firmly in the hands of economic egalitarianism. Currently, Korea has adopted increasingly egalitarian economic institutions and policies under the populist pressure for "economic democracy", which has survived for a quarter of a century and has gained even more steam recently, thereby shaping economic policy to become more egalitarian since the post-President Park era up until the present day. Consequently, Korea's "selection and concentration" policy that was a decisive factor

• TP Distribution Map

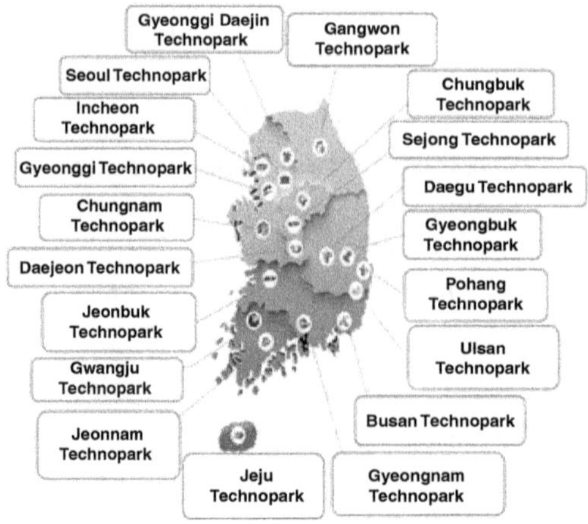

• Yearly Status of each TP Designation & Establishment

Period	Province	Year
Stage 1 (Exemplar 6 TPs)	Incheon, Gyeonggi, Daegu, Gyeongbuk, Gwangju, Chungnam	1997. 12
Stage 2 (Later 8 TPs)	Busan, Pohang	2000. 12
	Chungbuk, Jeonbuk, Jeonnam, Gangwon	2003.12
	Gyeongnam, Ulsan	2004. 12
Stage 2-2 Since 2005	Gyeonggi, Daegu	2005. 3
	Seoul	2005. 9
	Daejeon	2008. 1
	Jeju	2010. 1
	Sejong	2018. 11

Figure 8.2. Korea technopark distribution map and information for their establishment year.

for the Korean miracle in the developmental era has been substituted by the "balanced development" mentality.

Therefore, to strengthen and support SMEs and regionally balance development, as a major part of the Technopark, although the Korean government designated six technoparks in 1998 in the first place, it later established two more in 2000, four more in 2003, another two in 2004, has added five more since 2005 and attempted to promote technoparks. Therefore, 19 technoparks are in operation in 17 metropolitan cities and provinces across Korea as of 2020 (Figure 8.2). The map shown in Figure 8.2 also shows each technopark's specialized industries.

8.2 Major Roles and Functions of the Korean Technoparks and its Evolution

8.2.1 *Major roles and functions of technoparks in Korea*

There used to be two main streams of theories regarding science and technology parks—(1) the economic geography perspective and (2) the institutional perspective. From the economic geography perspective, a science park and its surrounding region are regarded as an entity consisting of specialized firms with an evolving structure of inter-firm linkages and agglomerative effects. The roles and contributions of science parks are examined in the context of regional development or regional innovation systems [e.g., Markusen, 1996; Storper and Harrison, 1991].

On the other hand, from the institutional perspective, a science and technology park is considered as an institution providing assistance for its tenants in specific policy-based or mechanism-based ways. This view emphasizes issues such as the functioning of incubators, the degree of spin-offs and whether STPs confer competitive advantages to their tenant firms as well as positive spillover effects to firms located in their complex and their regional economy. The tendency to view and promote STPs as a specialized infrastructure to house technology-based firms has focused particularly on the direct and tangible contributions of STPs and the institutions and mechanisms within them. The contributions encompass job creation and the quality of employment, the contribution to R&D investment and output, the amount of venture capital raised, as well as the role of universities within the science parks.

However, those two streams of theories could not be sufficient to explain the role and function of technoparks because they have been asked to play the role of a platform enabling value-creating interactions between producers and consumers or service providers and recently users. Especially, as we entered the so-called 4th Industrial Revolution era, platform business has drawn so much attention that technoparks are asked to evolve or be transformed to platform. Therefore, we have to add one more theory of platform to explain STPs more comprehensively.

In fact, Korean technoparks originated as an institution that focused on establishing infrastructures, incubating high-tech firms, and promoting university–industry–government (UIG) relations (see Figure 8.3). Actually, it has been supplementing the limitations of institutional systems of UIG (university–industry–government) relations by coordinating and cooperating with each other.

However, after President Moo-Hyun Roh assumed power in 2003, by focusing on distributive policies, the government planned to relieve

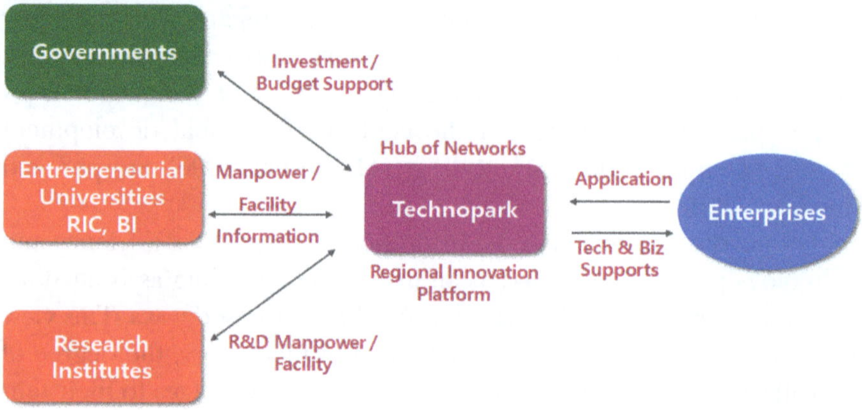

Figure 8.3. Technopark as a UIG facilitator.
Source: Excerpt from Rhee [2019a].

the unbalanced development between regions and set up a new political agenda—the so-called entire country's balanced development between regions through devolution and innovation. Thus, it attempted to foster technoparks as an important means for regional technology innovation and economic development [Lee *et al.*, 2004]. Since then, technoparks have been working as catalysts for regional economic development or revitalization and to promote economic growth [Rhee, 2016].

Since 2013, Korean technoparks have been asked to play as comprehensive SMEs supporters in each province—that is a kind of hub of networks, since 19 technoparks in Korea have grown into a hub of business support institutions in the last 22 years. Of course, to accelerate innovative growth in regions, technoparks lead the convergence of technologies, and strengthen networking and collaboration with the Centers for Creative Economy and Innovation and other regional supporting or innovative agencies. Now, they have evolved to platform.

In sum, technoparks in Korea have gone through four stages of development—firstly, space creation (for corporate innovation) and business incubators within the complex; secondly, UIG facilitation; thirdly, constitution of organization, that is, vision provider or regional industry promotion policymaker for regional development through fostering strategic industries [Rhee, 2016]; and fourthly, regional integrated business platform as a hub of network for business ecosystems across provinces. However, the first three stages have not taken place sequentially or independently, but concurrently, leading to an under-one-roof total business support system; that is, a technologically intensive industry park that puts every essential function in a single spot for effective high-tech industry development as well as a 'UIG' facilitator and policymaker and later, platform. Put more specifically, technoparks have established the "Corporate–Support Platform" which provides a one-stop service including initiating the business idea, pioneering the market, and even exporting products [Rhee, 2019a]. Then, they have evolved to platforms based on enabling value-creating interactions between startups and SMEs, large firms, service providers, and users [Rhee, 2019a]. Here, the producers or providers are the "new" startups or SMEs, while the buyers or users are the big and large firms, hidden champions or other SMEs. Therefore, Korean technoparks perform or play all four stages or functions at the same time [Rhee, 2019a] (Figure 8.4).

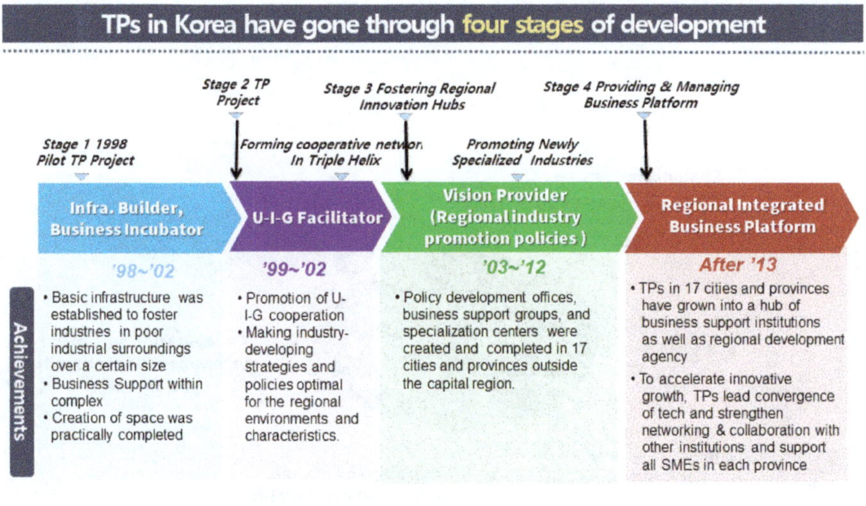

Figure 8.4. Evolutionary stages of Korea technopark.
Source: Excerpt from Rhee [2019a].

8.2.2 *The evolution of Korean technopark platform role*

As previously mentioned, technoparks have recently been evolving into the stage of a 'Business Platform'. According to Van Alstyne *et al.* [2016], "a platform provides an open, participative infrastructure and rules for a marketplace that brings together producers and consumers". That is, a platform provides an infrastructure for these interactions and sets governance conditions for them. The platform's purposes are to consummate matches among users and facilitate the exchange of goods, services, or some sort of social currency, thus enabling meaningful value exchanges between all participants [Van Alstyne *et al.* 2016] (see Figures 8.5–8.7).

There are four types of major players in the ecosystem which fill four main roles but may shift rapidly from one role to another—owners control their intellectual property and governance (e.g., Google owns Android); providers serve as the platforms' interface with users (e.g., apps on Android); producers create their offerings (e.g., mobile devices are providers on Android); and consumers use those offerings (e.g., buyers or users of the offerings) [Van Alstyne *et al.*, 2016]. Understanding the relationships both within and outside the ecosystem is central to platform strategy.

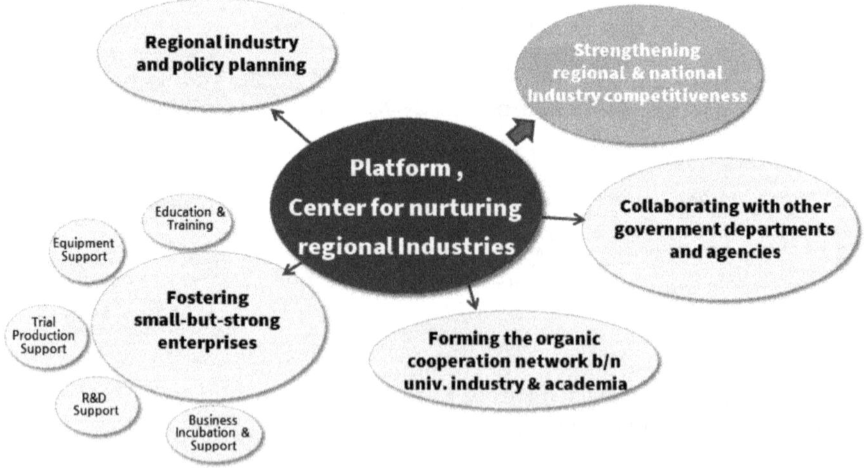

Figure 8.5. Technopark as platform of center for nurturing regional industries.
Source: Excerpt from Rhee [2019a].

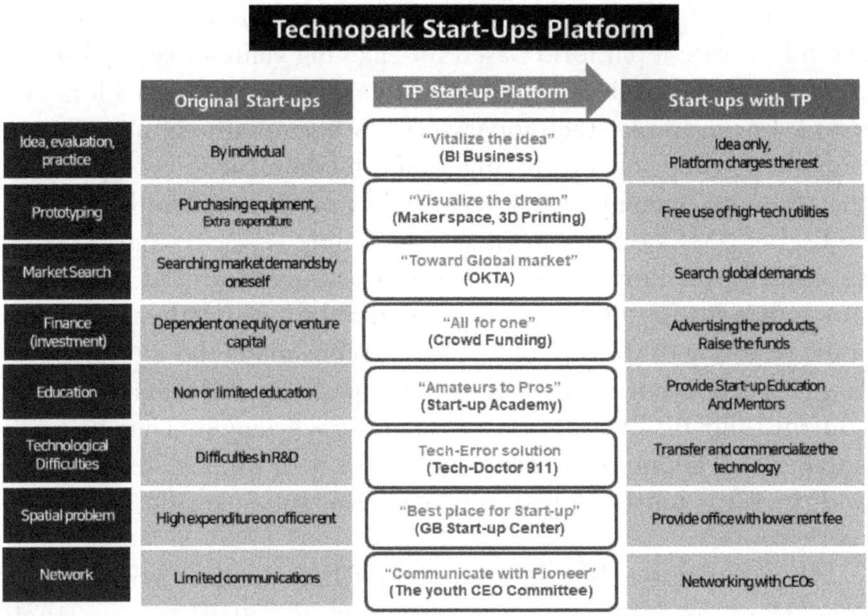

Figure 8.6. Technopark as platform of start-ups.
Source: Excerpt from Rhee [2019a].

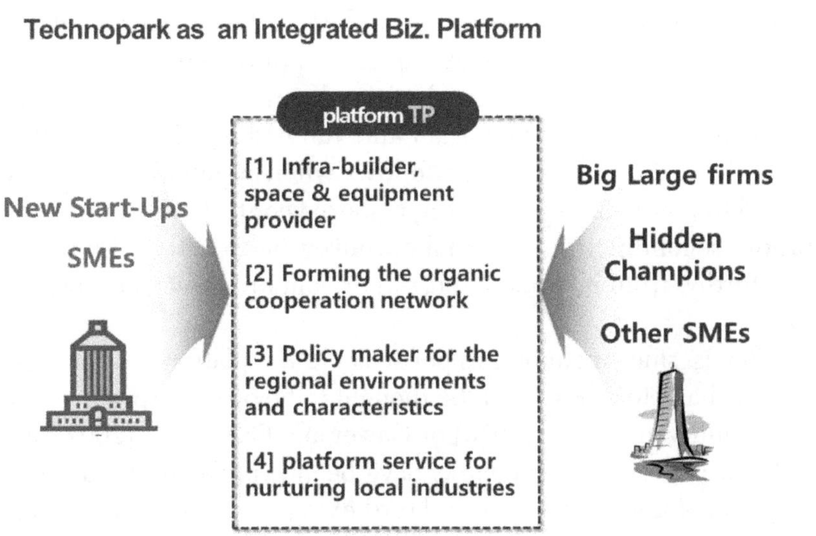

Figure 8.7. Technopark as platform of value-creating interactions between providers and users.
Source: Excerpt from Rhee [2019a].

According to Van Alstyne *et al.* [2016]'s argument, technoparks should provide a platform based on enabling value-creating interactions between producers and consumers or service providers and users. For example, technoparks act as both platform owners and providers, while startups or small- and medium-sized enterprises in the technopark complex are regarded as producers, and large firms and other SMEs outside technopark complex are consumers. More specifically, technoparks as platform providers can build infrastructure and rules for a transaction place that brings together producers (here referring to startups or SMEs) and consumers (here referring to large firms or other SMEs). Also, technoparks as platform owners can control their intellectual property and governance. That is, technoparks can define who participates, how it divides or shares value, and, perhaps most importantly, how to resolve conflicts between participants.

However, it may be excessive to apply to technoparks the concept of platform as defined by Van Alstyne *et al.* [2016] because their concept came from private sectors. Nonetheless, if we borrow some part of Gawer and Cusumano's "Platform Leadership" [2002], it can be complementary with the concept defined by Van Alstyne *et al.* [2016] because it is based on the industrial level. Various platforms have existed for years. For example, shopping malls link consumers and merchants, and recent platform businesses are Apple, Uber, Airbnb and Alibaba. Also, Parker and Van Alstyne [2016] emphasized those three characteristics as follows when pipeline businesses are shifted to platform—firstly, from resource control to resource orchestration; secondly, from internal optimization to external interaction; and thirdly, from a focus on customer value to a focus on ecosystem value.

That is, since technoparks per se act as the platform of quasi-public sector's characteristics, it can be thought to be overinterpreting to apply Parker and Van Alstyne [2016] or Gawer and Cusumano [2002] directly to them. However, technoparks have been evolving into platforms for Corporate–Support Platform and further to platforms based on enabling value-creating interactions between startups and SMEs and other SMEs and large firms and service providers and users. In other words,

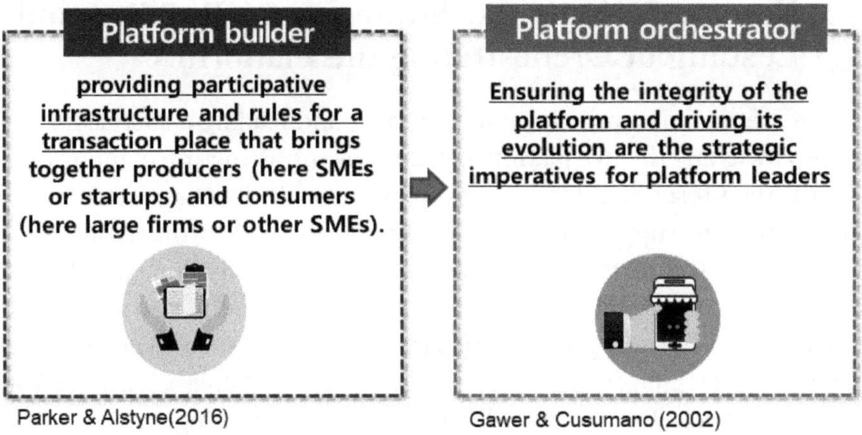

Figure 8.8. Technopark as platform of both builder and leader or orchestrator.
Source: Excerpt from Rhee [2019b].

technoparks have already established platforms where providers and users interact with each other.

Therefore, it is very important to build the platforms and sustain and develop them effectively to increase or add value of participants. Thus, if we borrow ideas of those models and converge them, they may be very useful in explanation [Rhee, 2019b]. For example, we can conceptualize 'platform building' by borrowing Parker and Van Alstyne [2016]'s platform provider and 'platform leading or orchestrating' by using Gawer and Cusumano [2002]'s industry level platform concept (see Figure 8.8).

As previously mentioned, according to Parker and Van Alstyne [2016], platform providers serve as the platforms' interface with users. On the other hand, according to Gawer and Cusumano [2002], platform leaders are companies who drive innovation in their industry. Ensuring the integrity of the platform and driving its evolution are the strategic imperatives for platform leaders. If they emerge and work with the companies supplying complementary products and services, they can together form an "ecosystem" of innovation that can greatly increase the value of their innovations as more users adopt the platform and its complements.

8.3 Korea's Technoparks Strategies for Building and Leading or Orchestrating the Platform

There are common principles for building and leading a successful science park as an effective platform; that is, the infrastructure of innovation such as the presence of clear vision and philosophy for the science parks, the continuous support for the expansion of networks from a long-term perspective, the proactive commitment and participation based on a trust and autonomy, the interactive learning through the sharing of relational knowledge, and building the virtuous circle of success and eventually platform which can be used in creating a superlative science park in other countries. In other words, technopark as an integrated business platform performs the following functions—as a space and equipment provider; forming the organic cooperation network; as policymaker for the regional environments and characteristics; and as platform service for nurturing local industries. In order to perform these functions, the technoparks should be transformed into effective platforms. Then, it is essential to build and lead or orchestrate the platform effectively.

8.3.1 *How to build the platform?*

The way to build the technopark as an effective platform is to provide participative infrastructure and rules for a transaction place that brings together producers (here referring to SMEs or startups) and users or consumers (here referring to large firms or other SMEs) because platform providers should serve as the platforms' interface with users. Therefore, Parker *et al.* [2016] argued that platforms must perform the following three key functions in order to encourage a high volume of valuable core interactions, which are essential to a successful platform and must be balanced—pull, facilitate, and match (see Figure 8.9).

8.3.1.1 *Pulling function*

Firstly, in order to pull both producers and buyers, the platform should propose clear values. Since the network effect is dependent, basically, on the number of users of a platform, it needs to attract as many users

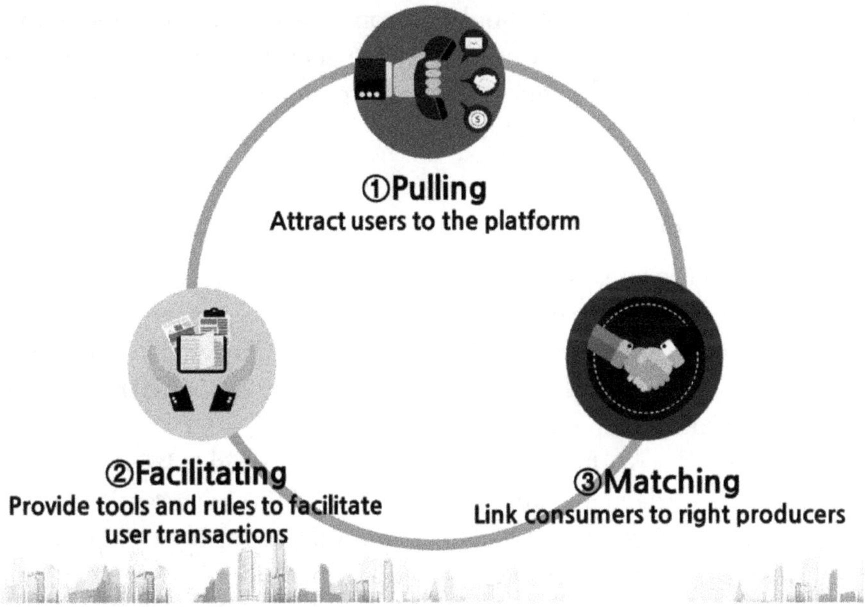

Figure 8.9. Three core functions for building effective platform.
Source: Excerpt from Parker *et al.* [2016].

as possible by providing the right value proposition or vision to each potential participating group.

Secondly, in order to build trust relationship, the platform must make users trust the, by showing evidence; not just telling, but also following through with their real actions.

8.3.1.2 *Facilitating function*

Firstly, in facilitating the interactions, the platform should provide tools and rules. If so, the tools can make it easy for users to interact with other members of the platform, which will make the system run more efficiently; i.e., they can help users communicate or interact with others in the platform. Rules, however, must be applied to control possible problems because all the frequent interactions among users do not guarantee the value of each transaction.

Secondly, the platform should be composed of type of modularity. Then, platform can grow or evolve as time goes on; therefore, the tools and rules need to be designed as a separate module following the same design rules and standard interfaces.

8.3.1.3 *Matching function*

Firstly, the platform should have quality producers since the quality of producers is critical to a platform. For example, a top-class record label on the iTunes platform could attract many music listeners. Therefore, we need to make an effort to keep quality startups or SMEs, or key persons or providers in our platform.

Secondly, the platform should link right users right. In order to do so, by building a database, all the information of producers and consumers must be managed thoroughly and carefully. Also, by setting procedures, a quick, easy and perfect way to match users is one of the important performance factors of a platform.

8.3.2 *How to lead or orchestrate the platform?*

As previously mentioned, platform leaders are companies who drive innovation in their industry. Therefore, in order to lead or orchestrate platform effectively, ensuring the integrity of the platform and driving its evolution are the strategic imperatives for platform leaders. If they emerge and work with the companies supplying complementary products and services, they can together form an "ecosystem" of innovation that can greatly increase the value of their innovations, as more users adopt the platform and its complements. Therefore, based on the case analysis of the Gyeonbuk Technopark, one of the first six designated and well-established technoparks, we derived three core approaches for leading or orchestrating platform in order to sustain and develop it effectively, as follows [Rhee, 2019a] (see Figure 8.10).

8.3.2.1 *The database approach*

In order to collect information of local firms in detail, a database should include the firm's name, revenue, address, employee

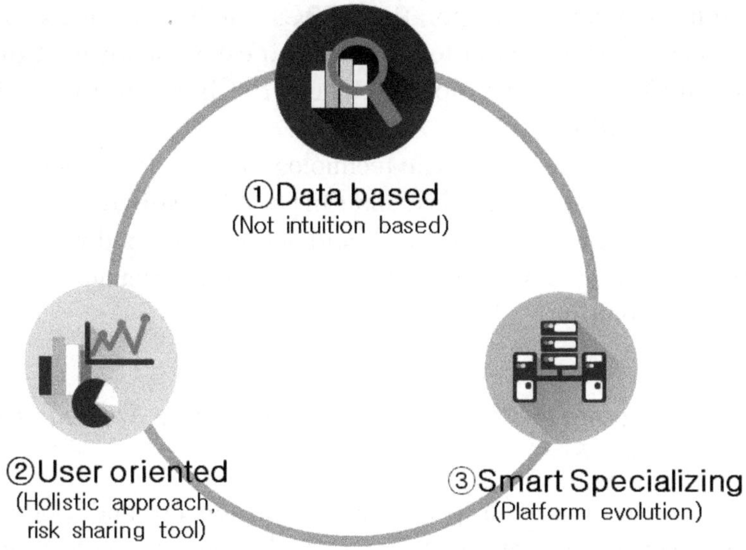

Figure 8.10. Three core approaches for leading or orchestrating effective platform.
Source: Excerpt from Rhee [2019a].

numbers, equipment, business transaction records, etc. In addition, the database should also be built—follow-up database of tenants and local firms such as the track records of firms' contact activities, performance evaluation of SMEs that received government support programs and human resource database such as the directory of technology experts, a map of business relationships at an individual level, etc.

8.3.2.2 *The user-oriented approach*

The platform should be both user-friendly by making itself holistic and a one-stop shop and an under-one-roof system, and provide risk-free stability by using Blockchain technology and act as a test bed.

Firstly, a holistic collaboration support system should be established for local firms and tenants, and total solutions should be provided by placing or clustering within technoparks all necessary agencies such as the Technology Licensing Office, the Korean Small and medium Business Corporation (SBC), the Technical Assurance Fund (KIBO), the Technology and Business Consulting Contact Center, the Intellectual Property Center, the Institute for Regional Program Evaluation, etc.

Secondly, by using blockchain technology, the platform should guarantee privacy and confidentiality, and thus build trust between producers and users. Also, the platform should act as a test bed which deals with the hyper-uncertainty of technology and market changes in the 4th Industrial Revolution.

8.3.2.3 *Platform evolution by adopting the 'Smart Specialization' approach*

Smart specialization is a place-based, partnership and bottom-up approach conceived within the reformed Cohesion policy of the European Commission and characterized by the identification of strategic areas for intervention based both on the analysis of the strengths and potential of the economy and on an Entrepreneurial Discovery Process (EDP) with wide stakeholder involvement. It is an innovative approach by enabling each province or municipality to identify and develop its own competitive advantages. In doing so, it can bring together local authorities, academia, business spheres and the civil society, working for the implementation of long-term growth strategies supported by central government funds.

Starting point of Smart Specialization should be to find out "What Have We Done?, Where Are We Now?, What Do We Have?, Then Where Are We Going?" In addition, basic philosophy of regional development paradigm in combination with smart specialization should be based not on "Wish-to-do Projects" (which could be performed by central governments and later designated to regional governments), but on both "Can-do Projects" (which can be performed by regional governments following market principle) and "Must-do Projects" (which should be performed by regional governments and supported by central governments).

8.4 Major Performances of Technoparks and Reasons for their Success

The Korean Technopark Project has been regarded as one of the best and rarest among various regional projects in terms of the success of technopark itself, progress of tenant company support, regional SMEs support, regional industrial promotion planning and outcome, etc.

8.4.1 *Major reasons for their success*

8.4.1.1 *Technopark-related statistics until 2019*

As of 2019, the total number of technopark staff (or employees) has risen to 2,640 and the annual budget of the 19 technoparks amounts to $13,889 million (see Figure 8.11).

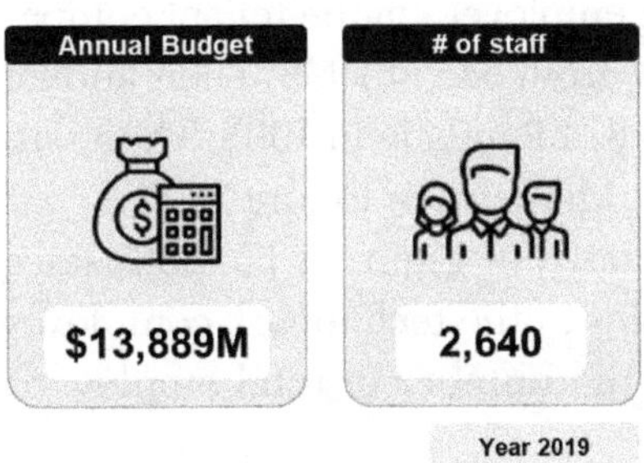

TP	Budget	Staff
Seoul TP	208	52
Busan TP	1,008	201
DGTP	1,349	216
ICTP	1,374	270
GJTP	394	95
DJTP	776	141
USTP	805	171
SJTP	105	43
GGTP	498	150
GWTP	498	92
CBTP	870	150
CNTP	906	132
JBTP	631	140
JNTP	782	149
GBTP	803	160
GNTP	1,367	202
JJTP	830	142
GGDJTP	320	71
PHTP	365	63
Total	**13,889**	**2,640**

Figure 8.11. Technoparks' annual budget and number of staffs.

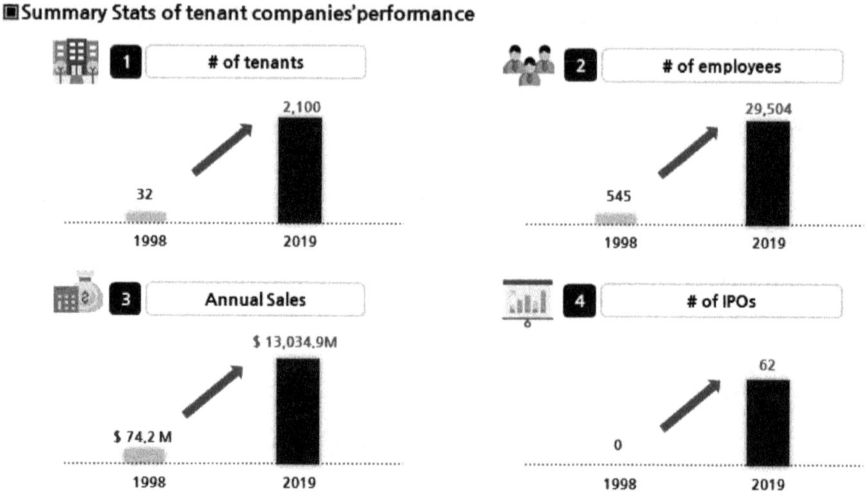

Figure 8.12. Summary stats of tenant companies' performance.

8.4.1.2 *Tenant company-related statistics until 2019*

As of 2019, the total number of tenant companies are 2,100, compared to only 32 in 1998, and the total number of employees in the tenant companies has also sharply increased to 29,504, from 545 in 1998. Their annual sales amount to $13 billion compared to $74.2 million in 1998. There are 62 companies listed in the Korean Stock Market (see Figure 8.12).

As seen in the tenant companies' status in Figure 8.13, both sales volumes and number of tenant companies in the technopark complexes have grown steadily since 2011, while the number of total employees remain stagnant.

8.4.1.3 *Regional SMEs support-related statistics until 2017*

More than half of regional SMEs with more than 10 employees have been beneficiaries of technopark business support services including R&D, non-R&D (e.g., consulting services, funding, marketing services), and other business supports. In addition, 18 technoparks hold 2,899 units of equipment (higher than $30,000 per unit). Up to now, all 18

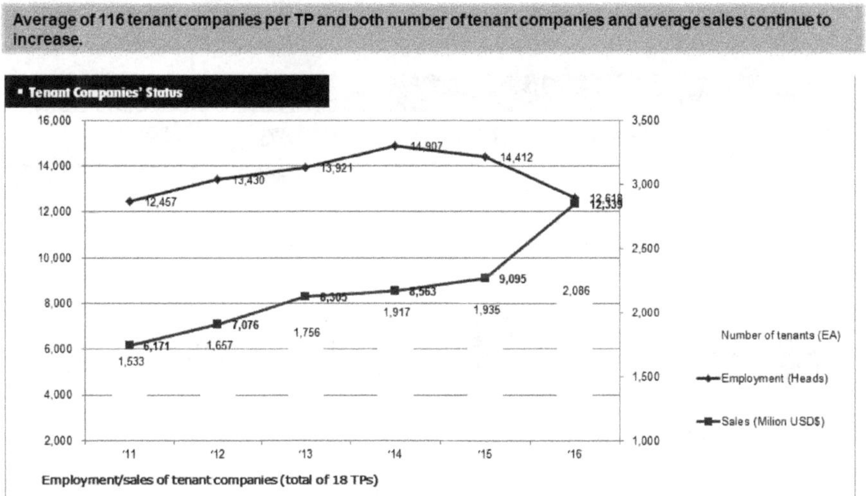

Figure 8.13. Tenant companies' sales and number of employees.

technoparks have spent almost $648,805,000 for those equipment.[1] One of important services the technoparks provide is the equipment rental service. More than half of the equipment cost less than $100,000 while 116 are more expensive than $1 million. Half of them are less than 10 years and older than five years.

While each technopark holds an average of 128 kinds and 166 pieces of equipment, it collects $1.31 million on average as equipment utilization fees a year. The rate of operation of the main equipment purchased over $30,000 in the last five years recorded 52.5% (see Figure 8.14). Such services of equipment rent have been found to be very instrumental and beneficial to regional SMEs. Through these various services to the regional SMEs as well as tenant companies, technoparks have become a hub of business support institutions or networks, and recognized as one of the best regional projects in Korea.

[1] Until June 2018, there were 18 technoparks because Sejong Technopark was established in 2019. Therefore, the data on only 18 technoparks were available.

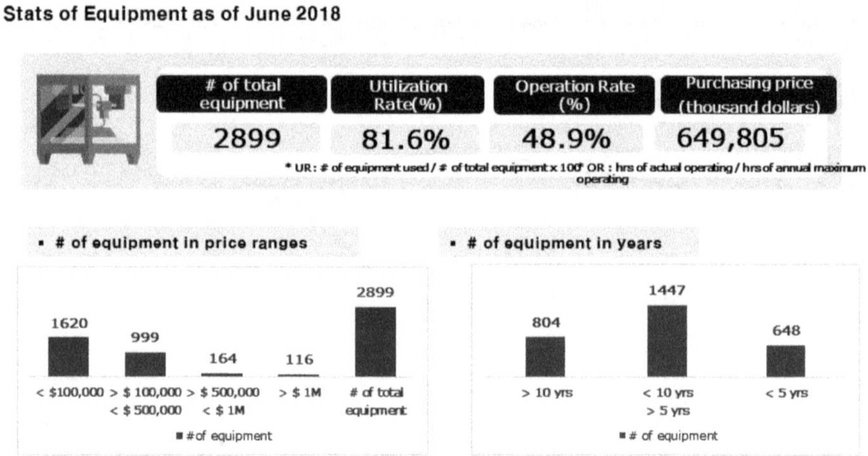

Figure 8.14. Statistics of equipment in technopark.

8.4.1.4 *Regional industrial promotion planning and outcome related statistics*

Korean technoparks have led regional development as frontrunners with various regional industries promoting policies since 1998. Specifically, since 2003 when former President Moo-Hyun Roh assumed power, by focusing on distributive policies, he asked the local governments to initiate their own policies in a bottom-up style. Therefore, local or regional governments started to rely on technoparks for their regional industrial promotional policies and even economic plans. In addition, the Korean central government also attempted to foster technoparks as an important means for regional technology innovation and economic development [Lee *et al.*, 2004]. Since then, technoparks have been working as catalysts for regional economic development or revitalization and promoting economic growth—that is, the Korean version of regional development agencies.

For example, through the project for cultivating regional strategic industries (2002–2012) which lasted during those periods, the GRDP of non-capital regions has increased by an average annual growth rate

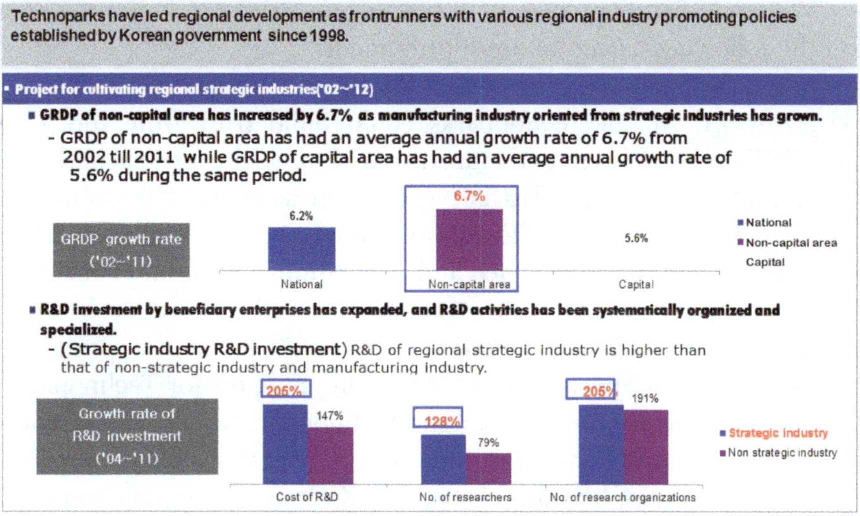

Figure 8.15. Technoparks' performance comparison in growth rate and R&D investment between non-capital region and capital region.

of 6.7% from 2002–2011, as the manufacturing industry oriented from strategic industries has grown, while the GRDP of the capital region has had an average annual growth rate of 5.6% during the same period. In addition, R&D investment by beneficiary enterprises has expanded and R&D activities has been systematically organized and specialized. For example, the R&D of the regional strategic industry was higher than that of non-strategic industry and manufacturing industry (see Figure 8.15).

8.4.2 *Major reasons for their success*

Without any doubt, we can acknowledge the significant contributions made by technoparks. Korean technoparks could not be the same as others or as existing successful science parks in developed countries in many ways because each technopark has its unique conditions of specialized industries, history and area, based on industrial preferences.

We could list many factors in the technoparks' successes in Korea but the following six are the most prominent:

(1) *Trust*: Above all, technoparks could build trust from the local SMEs, government, and citizens based on quality service and core competences. In the last 20 years, each technopark has obtained trust from the local SMEs, governments, and citizens by providing and showing quality service and core competences, and contributing to each province.
(2) *Legislation*: A special legislation was enacted; that is, the special law titled 'Act on Special Cases Concerning Support for Technoparks' was enacted in 1998 for Korean technoparks which have guaranteed the stability and consistency of the operations of technoparks. In other words, because of that law, Korean technoparks were not affected by regime changes and could get the initially guaranteed funding from the government and sustain their institutions.
(3) *Governance*: Good governance between the Central and local governments made the support system more stable.
(4) *Strategic industries*: Technoparks have been supported with the help of the central government and their regional strategic industries that could have a competitive edge in order to develop regions in a balanced way.
(5) *Systemization*: Technoparks have provided a systemic model for company support and industry promotion programs based on competition.
(6) *Collaboration*: Technoparks have supported small- and medium-sized enterprises via collaborations with other regional business support institutions or centers (for example, Centers for Creative Economy and Innovation, Korea SMEs and Startups Agency, KIBO, Intellectual Property Centers, etc.).

In sum, technoparks will go forward with their endogenous development based on the innovative economy and establish the integrated support systems customizing corporate demands. They have been and will continue to be promoted as an Integrated Business platform builder and orchestrator of cooperation between industry, university, research institution and government, supporting local small but strong businesses, strengthening their global competitiveness through smart specialization and professionalism in local specialized industries, and driving innovative growth and regional development, and local citizens' happiness by supporting businesses and creating jobs in the region. Korean technoparks have already transferred their experience to those countries which wanted to reduce regional gap by developing regional industries.

8.5 Future Technopark Model as a Platform

As previously mentioned, technoparks have been evolving into platforms for Corporate–Support Platform and further into platforms based on enabling value-creating interactions between startups and SMEs and other SMEs and large firms and service providers and users. In other words, technoparks have already established platforms where providers and users interact with each other. Therefore, it is very important to build the platforms, and sustain and develop them effectively to increase or add value of participants.

However, although they have been successfully transformed to platform service organizations, they are still hardware oriented—namely, hardware facilities and production of prototypes, etc. For example, though some of them are successful platforms, most of them are focused on post-business incubation facilities, test and measurement equipment (established technology), and fragmented data handling, etc.

Therefore, in order to truly become platforms, that is, to-be service platforms, they should be focused on innovative SME support programs and database (DB) analysis for industrial strategies. For example, they should be equipped with life cycle, packaged service platforms for SMEs scale-up (venture capital, accelerator), smart center for 4th Industrial Revolution (pilot line testing, simulation, consulting, etc.), and Technopark Knowledge Center (DB analysis, publishing, education, etc.). In other words, the technopark should be a 'Technopark Service Platform' (see Figure 8.16). It should be focused on the three functions, as follows [Rhee, 2019b]:

(i) Technopark Knowledge Center
Evidence-based, industrial policy planning
New services based on firm DB and know-how (publication, education, consulting, etc.)

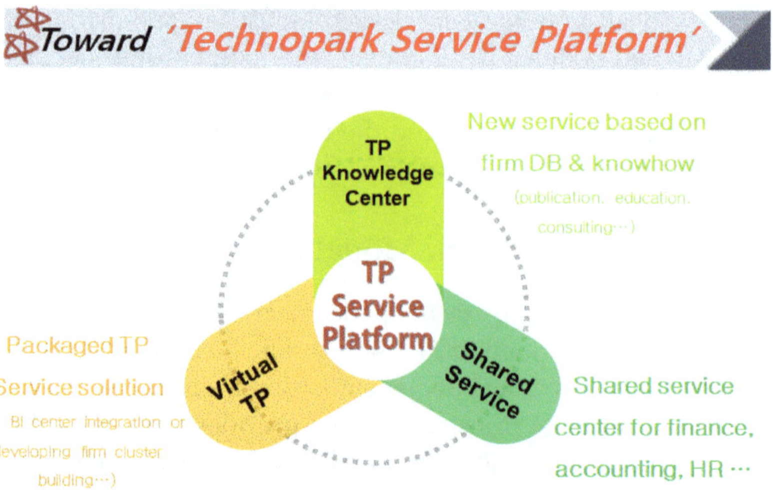

Figure 8.16. To-be technopark model as a service platform.
Source: Excerpt from Rhee [2019b].

(ii) Virtual Technopark
Packaged Technopark Service solution (i.e., Business Intelligence (BI) center integration or developing firm cluster building…)
Smart Manufacturing Innovation Center (interoperability test bed service, technical verification…)

(iii) Shared Service
Shared service center for finance, accounting, HR, etc.

All around services for New Technology-Based Firms (NTBF), business scale-up (venture capital, large-scale manufacturing, marketing, etc.).

References

Choudary, S.P. (2016). *How to Build a Successful Platform Business*, Blog Posts/INSEAD Blog.

Gawer, A., and Cusumano, M.A. (2002). *Platform Leadership*. Boston: Harvard Business School Press.

Gawer, A., and Cusumano, M.A. (2008). *Platform Leaders*. MIT, Sloan Management Review, pp. 68–75.

Gawer, A., and Cusumano, M.A. (2014). Industry platforms and ecosystem innovation. *Journal of Product Innovation Management*, *31*(3), 417–433.

Jwa, S.-H. (2017). *The Rise and Fall of Korea's Economic Development: Lessons for Developing and Developed Economies*. Palgrave Macmillan.

KOICA and UNESCO. (2018). Training for Trainers on Development of Science Parks in Indonesia, April 2018. Jakarta: Guide Book.

Lee, S.K., Park, S.C., and Lee, K.R. (2004). Build-up of regional innovation systems and roles of Technoparks, *National Territorial Planning*, *39*(2), 255–270 (in Korean).

Markusen, A. (1996). Sticky places in slippery places: A typology of industrial districts, *Economic Geography*, *72*(3), 283–313.

Parker, G., Van Alstyne, M.W., and Choudhary, S.P. (2016). Platform Revolution: How networked markets are transforming the economy and how to make them work for you, W.W. Norton & Company.

Rhee, J. (2016). *Evolution of Technoparks: An instance towards Regional Boost for Developing Countries: Experience from Korean Technopark, 2016 Proceedings*. Daejeon, Korea: World Technopolis Association.

Rhee, J. (2019a). *Korea TechnoPark: An Integrated Business Platform, 2019 Proceedings*. Daejeon, Korea: World Technopolis Association.

Rhee, J. (2019b). *TechnoPark as an Integrated Business Platform: Present & Future, Presented at the Plenary Session of the 23rd Annual Conference of Asia Science Parks Association*, Hsinchu, Taiwan.

Shin, D.-H. (2001). An alternative approach to developing science parks: A case study from Korea. *Papers, Regional Science*, *80*, 103–111.

Storper, M., and Harrison, B. (1991). Flexibility, hierarchy and regional development: The changing structure of industrial production systems and their forms of governance in the 1990s. *Research Policy*, *20*, 407–422.

Van Alstyne, M.W., Parker, G., and Choudhary, S.P. (2016). Pipelines, platforms, and the new rules of strategy. *Harvard Business Review*, April 2–9.

© 2023 World Scientific Publishing Company
https://doi.org/10.1142/9789811257186_0009

CHAPTER 9

Impacts of the First to the Second Generation of Smart City Initiatives in Japan

Jun Yamashita

Social and Cultural Studies, Kyushu University, Fukuoka, Japan
Email: yamasita@scs.kyushu-u.ac.jp

9.1 Introduction

With the rapid development of information and communication technologies (ICTs), including 5G technology, a number of smart city policies have been planned and implemented in recent years all over the world. While the implementations of these smart city policies have been summarized, these policies have also been classified in terms of their implementation [Alawadhi *et al.*, 2012; Angelidou, 2014; Marsal-Llacuna *et al.*, 2015; Kourtit *et al.*, 2017; Viale Pereira *et al.*, 2017; Yigitcanlar and Kamruzzaman, 2018]. Previous studies addressing such summaries and classifications of policy implementation for smart cities could be regarded as the policy research for the stage of policy implementation, namely the 'Do' stage of the policy cycle (the Plan-Do-Check-Act cycle, or PDCA cycle). On the other hand, there are relatively few studies on the evaluation stage, also known as the 'Check' stage, of the PDCA cycle, especially on the impacts of smart city policies.

Among the few policy evaluation studies available on smart city policies, Carogliu *et al.* [2019] used patent data to reveal that smart city policies contributed to the growth of high-tech technologies in European Union (EU) cities, where smart city policies have been implemented, rather than other EU cities where smart city policies have not been implemented. Bower [2019] also showed that the smart city policies implemented in Istanbul had a positive impact on the following policy areas—infrastructure including transportation, human capacity building including education, safety and security, health services, and emergency management.

From the viewpoint of the policy cycle, it is necessary to disclose the impacts of the present smart city policies on future ones. However, policy evaluation studies on smart cities are not comprehensive in this study domain. Regarding the research objectives in the present chapter, namely smart city policies in Japan, impacts of the first generation of these policies on later ones have not yet been fully covered in either official documents or academic literature. Further accumulation of evaluation studies on smart city policies is, therefore, needed.

After various discussions on the aforementioned policy evaluations, policy studies turned their attention to conceptual frameworks for smart city policies from the viewpoint of the stage after the evaluation stage, called the 'Act' stage. The conceptual framework proposed by Chourabi *et al.* [2014] was a widely used one for smart city policies. They presented a two-layer model centered on smart city initiatives as the core. The three elements of technologies, organizations, and policies, which directly affect the success of smart city initiatives, are situated in the layer nearest to the core. On the other hand, the five elements of governance, people and communities, economy, built infrastructure, and natural environment, which are considered as less influential elements than the aforementioned three ones, are located in the outer layer of this conceptual framework.

Further expansions of Chourabi *et al.* [2014]'s conceptual framework were proposed. One direction of such expansions was the incorporation of stakeholders [Deakin, 2014; Etzkowitz and Zhou, 2006; Lombardi *et al.*, 2011]. In this direction, the Triple/Quadruple helix models were incorporated into Chourabi *et al.* [2014]'s framework for further articulation of the element "people and communities", which activates technologies. In addition to stakeholders, some policy frameworks address the importance of governance, which could integrate these stakeholders [Castelnovo *et al.*, 2015; Meijer and Bolivar, 2015]. In another direction apart from the stakeholders, some frameworks focused on further articulated determinants shown in the outer layer of Chourabi *et al.* [2014]'s framework [Fernández-Güell *et al.*, 2016; Monzon, 2015]. Although the two directions were proposed, each study dealt with only one of these articulated elements rather than the whole. Fernandez-Anes *et al.* [2018] exceptionally presented their conceptual framework covering the whole elements.

The framework presented by Fernandez-Anez *et al.* [2018] consists of three layers from the center to the outer layer—stakeholders, smart city dimensions, and global trends. In this framework, smart city policies are assumed to be decided and implemented by interactions among three layers and their elements. The inner layer is composed of four stakeholders—political, social, economic and knowledge stakeholders. The idea of these four stakeholders, namely government, citizens, industry and research institutes, relies on the Quadruple Helix Model for innovation [Carayannis and Campbell, 2009]. The middle layer, the smart city dimension, consists of six elements—governance, economy, environment, mobility, people and living. Giffinger *et al.* [2007] proposed these six elements, and this six-element model for smart cities is also employed in the European Parliament [2014]. Among the six elements, governance has the chief role to integrate the other five. Finally, the outer layer of global trends comprises six areas—climate change, social polarization, new governance models, global urbanization, economic instability, and technological innovations. Although this framework consists of the aforementioned three layers, each element on these layers is inter-related. Thus, stakeholders respond to challenges caused by global trends, and prioritize and implement smart city initiatives in the six smart city dimensions. However, Fernandez-Anes *et al.* [2018]'s framework was validated only for smart city policies in Vienna. It is, therefore, necessary to evaluate the validation of this framework by applying this framework to smart city policies in other cities.

The purpose of the present chapter is to reveal the present situation of the latest smart city policies, including the impacts of the first generation on these states of the art policies, using the conceptual framework presented by Fernandez-Anez *et al.* [2018]. The present chapter was also intended to evaluate the validity of this conceptual framework because it seems that no studies have tackled framework evaluation.

9.2 Smart City Policies in Japan

The national smart city policy started in 2008 in Japan. In this year, the Cabinet Office initiated the eco-model and future cities project on the basis of the "future city" initiative, while the Agency for Natural Resources and Energy, under the Ministry of Economy, Trade and Industry, launched the next generation energy and social system demonstration project, in 2010 (Table 9.1). Although both smart city national policies aim at the construction of smart sustainable cities, their directions are slightly different. While the former is tightly connected to the term "smart", namely information communication technology (ICT), the latter is oriented toward environmental sustainability. Two cities, Yokohama and Kitakyushu, were designated not only as model cities for the next generation energy and social system demonstration project, but also as eco-model and future cities.

In the eco-model and future city project, 23 cities were designated as eco-model cities in total (i.e., 13 cities in 2008, seven in 2012, and three in 2013) and 11 cities as future cities in 2011 (Figure 9.1). To extend the "FutureCity" initiatives, the Cabinet Office started the United Nations Sustainable Development Goals (SDGs), Future Urban project in 2018. In this project, the SDGs are utilized as a new tool to foster the creation of an economically and socially sustainable society across Japan.

Table 9.1. Smart city-related major legislation, projects and events in Japan.

2008	(P) Low-Carbon City Model Project (MLIT and MoE) (P) Ecological urban development project (MLIT) (P) Eco-model and future cities project (CO)
2010	(P) The next generation energy and social system demonstration project (METI)
2011	(P) FutureCities project (CO) (P) Project for Promoting Introduction of Smart Communities (METI)
2012	(L) Low Carbon City Promotion Act (P) Supportive project for building energy storage and saving models for towns, homes, and transportation (MLIT) (P) ICT Urban Development Promotion Project (MIAC)
2013	(P) Pilot project for building low-carbon urban development plans based on public participation (MoE)
2015	(P) Promotion project for creation of ICT towns, people and jobs (MIAC)
2016	(L) The 5th Science and Technology Basic Plan (L) Basic Act on the Advancement of Public and Private Sector Data Utilization (L) Comprehensive Strategy on Science, Technology and Innovation
2017	(P) Promotion project for data utilization driven smart cities (MIAC)
2018	(P) SDGs Future Cities (CO)
2019	(P) Smart city model project (MLIT) (C) G20 Ministerial Statement on Trade and Digital Economy (Tsukuba, Japan) (June)

Notes: (P): Projects, (L): Legislations and plans, (C): Conferences.
Affiliations: CO: Cabinet Office; METI: Ministry of Economy, Trade and Industry; MIAC: Ministry of Internal Affairs and Communications; MLIT: Ministry of Land, Infrastructure, Transport and Tourism; MoE: Ministry of the Environment.

Selected Cities and Regions

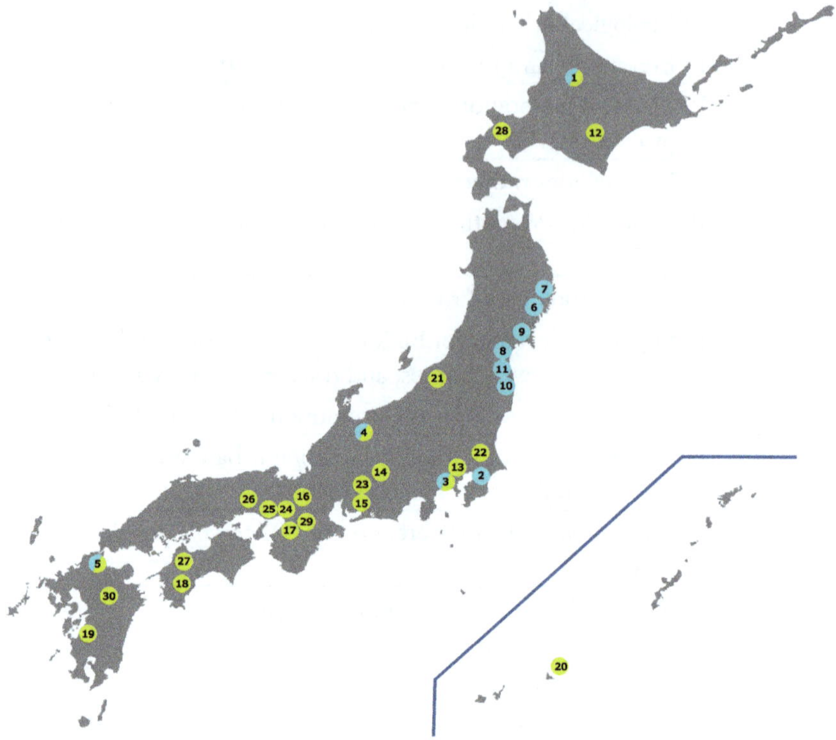

["FutureCity"]	[Eco-Model City] [Selected Cities in fiscal 2008]		[Selected Cities in fiscal 2012]	[Selected Cities in fiscal 2013]
1 Shimokawa Town (Hokkaido)	1 Shimokawa Town (Hokkaido)	19 Minamata City (Kumamoto Prefecture)	21 Niigata City (Niigata Prefecture)	28 Niseko Town (Hokkaido)
2 Kashiwa City (Chiba Prefecture)	12 Obihiro City (Hokkaido)	20 Miyakojima City (Okinawa Prefecture)	22 Tsukuba City (Ibaraki Prefecture)	29 Ikoma City (Nara Prefecture)
3 City of Yokohama (Kanagawa Prefecture)	13 Chiyoda City (Tokyo)		23 Mitake Town (Gifu Prefecture)	30 Oguni Town (Kumamoto Prefecture)
4 Toyama City (Toyama Prefecture)	3 City of Yokohama (Kanagawa Prefecture)		24 Amagasaki City (Hyogo Prefecture)	
5 City of Kitakyushu (Fukuoka Prefecture)	14 Iida City (Nagano Prefecture)		25 Kobe City (Hyogo Prefecture)	
6 Kesen Regional FutureCity (Iwate Prefecture)	4 Toyama City (Toyama Prefecture)		26 Nishiawakura Village (Okayama Prefecture)	
7 Kamaishi City (Iwate Prefecture)	15 Toyota City (Aichi Prefecture)		27 Matsuyama City (Ehime Prefecture)	
8 Iwanuma City (Miyagi Prefecture)	16 Kyoto City (Kyoto Prefecture)			
9 Higashimatsushima City (Miyagi Prefecture)	17 Sakai City (Osaka Prefecture)			
10 Minamisoma City (Fukushima Prefecture)	18 Yusuhara Town (Kochi Prefecture)			
11 Shinchi Town (Fukushima Prefecture)	5 City of Kitakyushu (Fukuoka Prefecture)			

Figure 9.1. Eco-model and future cities in Japan.
Source: Promotion Council for the "FutureCity" Initiative [2016].

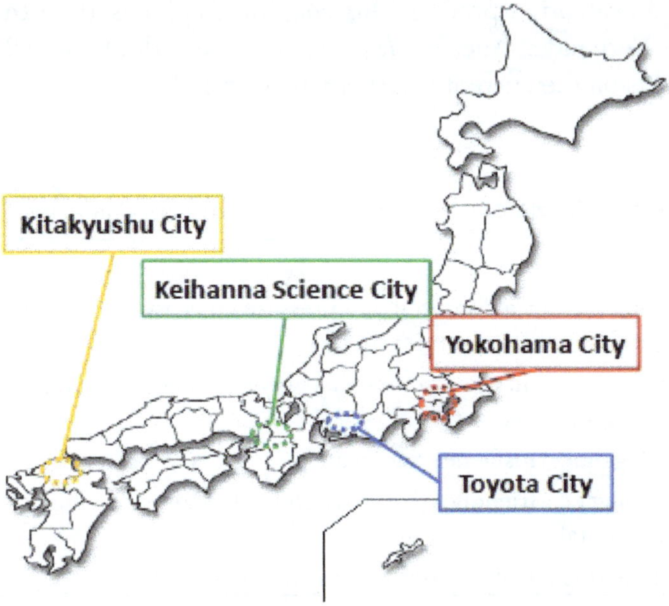

Figure 9.2. Four designated cities for the next generation energy and social system demonstration project in Japan.
Source: From Komiyama, Y. (2012) "Smart communities", as a presentation of US–Japan Renewable Energy Policy Business Roundtable 'New Opportunities for Renewable Energy in Japan & the United States'.

On the other hand, four cities—Yokohama, Toyota, Keihanna and Kitakyushu—were selected for experiments of the next generation energy and social system demonstration project (Figure 9.2). In this project, the smart community was defined as "a new form of social system that comprehensively manages the supply and demand of energy in the distributed energy systems, optimizes the use and application of energy, and incorporates lifestyle support services, including monitoring service for the elderly, through the energy management system utilizing IT and storage energy technologies, while making use of distributed energy resources such as renewable energy and cogeneration" [Japan Smart Community Alliance, 2015, p. 2].

In the evaluation process, this chapter also focused on the global trends of Fernandez-Anes *et al.* [2018]'s framework (Table 9.2), which had impacts on the current smart city policies in Japan.

Table 9.2. The global trends proposed by Fernandez-Anez *et al.* [2018].

1. Climate change
1.1. Reducing ecological footprint and pressure on ecosystems, promoting ecological functions of land
1.2. Increasing efficiency in resource management (energy, water, etc.) and promoting a circular economy
1.3. Fostering cities' resilience to climate change and disaster risks
1.4. Developing eco-friendly urban environments and responding to growing environmental concerns
1.5. Implementing a holistic approach to environmental issues
2. Social polarization
2.1. Promoting social inclusion, cohesion and equity
2.2. Enhancing the inclusion of migrants and refugees
2.3. Adapting the city's economic and social life to an aging population while attracting young people and children
2.4. Promoting equity in access to the labor market and the work-life balance
2.5. Eradicating spatial exclusion and promoting equity in access to housing and quality urban environments
2.6. Enhancing social diversity as a dynamic asset
3. New governance models
3.1. Changing to a more participative and inclusive democracy
3.2. Promoting citizenship via urban co-creation and co-management combining top-down and bottom-up models
3.3. Increasing the flexibility and resiliency of governance models
3.4. Improving the effectiveness of institutions, coordination among public bodies and multilevel governance (leading to more integrated sector policies)

Table 9.2. (*Continued*)

3.5. Incorporating and regulating innovative management systems at the local level while improving capacity building (i.e., PPP or PPPP, e-governance, etc.)
3.6. Enhancing territorial cohesion
4. Global urbanization
4.1. Managing the urban population growth while reducing negative externalities
4.2. Maintaining quality of life in cities, ensuring access to services in line with changes in demand (education, health, culture, safety, etc.)
4.3. Promoting interurban variety and cities' identity by protecting cultural heritage
4.4. Developing new planning tools for sustainable development (less urban sprawl, polycentric plans, increased density and diversity, mixed land use, urban refurbishment, etc.)
4.5. Fostering sustainable accessibility in cities and promoting sustainable, inclusive and healthy mobility when needed
4.6. Balancing urban growth and territorial development (managing the urban–rural balance)
5. Economic instability
5.1. Improving the resilience of economic systems and adaptation to changes in global and local economies
5.2. Improving the sustainability and diversity of local economies in balance with cities' specialization
5.3. Managing adaptation to innovation and knowledge-based economies while providing solutions to a broad skill base
5.4. Fostering human and social capital as source of innovation
5.4. Enhancing integration in global economies, promoting cooperation among cities and territories
5.5. Fostering employment creation with high quality standards
5.6. Achieving balance between competitiveness and quality of life
6. Technological innovations
6.1. Enhancing the adaptation of society, governance and economy to transformation through ICT

(*Continued*)

Table 9.2. (*Continued*)

6.2.	Coordinating new technologies for energy saving and reducing emissions through planning and governance tools
6.3.	Articulating mobility planning tools and policies with innovations in the sector
6.4.	Promoting technological innovation driven by social and human capital
6.5.	Reducing externalities in the implementation of new technologies (i.e., cybersecurity)

9.3 Policy Impacts

In this section, the current challenges Japan is facing are discussed, and the validity of Fernandez-Anez *et al.* [2018]'s conceptual framework for smart city policies is then examined. Next, Society 5.0, presented by the Cabinet Office, is briefly suggested for addressing these challenges. Finally, the latest smart city policies relying on the notion of Society 5.0 are shown. These policies include the establishment of smart city platforms.

9.3.1 *Global trends*

Japanese cities are characterized by the major challenges of the declining birth rate, an aging and shrinking population (Figure 9.3) and an increase in renewal and maintenance costs due to the aging of urban infrastructure (Figure 9.4). In the conceptual framework by Fernandez-Anez *et al.* [2018], the declining birth rate and an aging and shrinking population is referred to in subcategory 2.3, "Adapting the city's economic and social life to an aging population while attracting young people and children" within the category of "Social polarization", while the increase in renewal and maintenance costs due to the aging of urban infrastructure is mentioned in subcategory 4.2, "Maintaining quality of life in cities, ensuring access to services in line with changes in demand (education, health, culture, safety, etc.)" within the category of "Global urbanization" (Table 9.2).

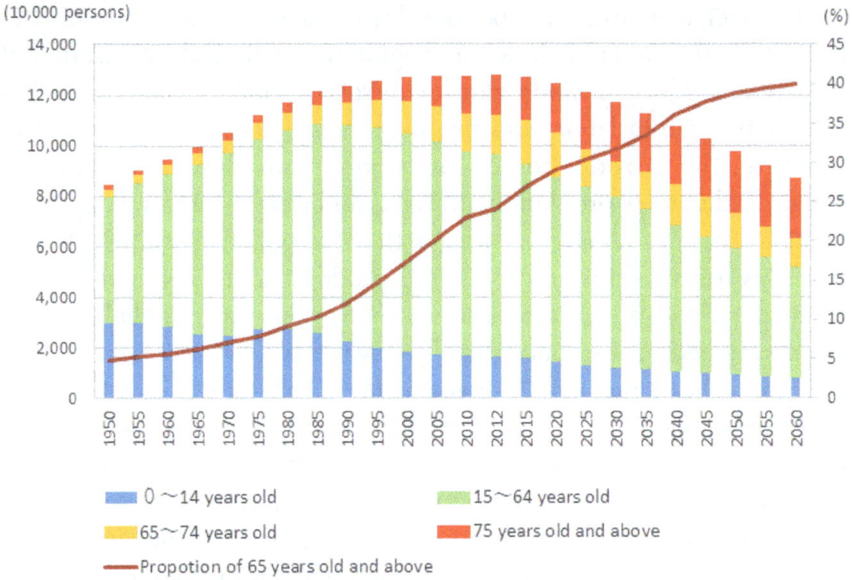

Figure 9.3. Population changes and estimation in Japan (1950–2060).

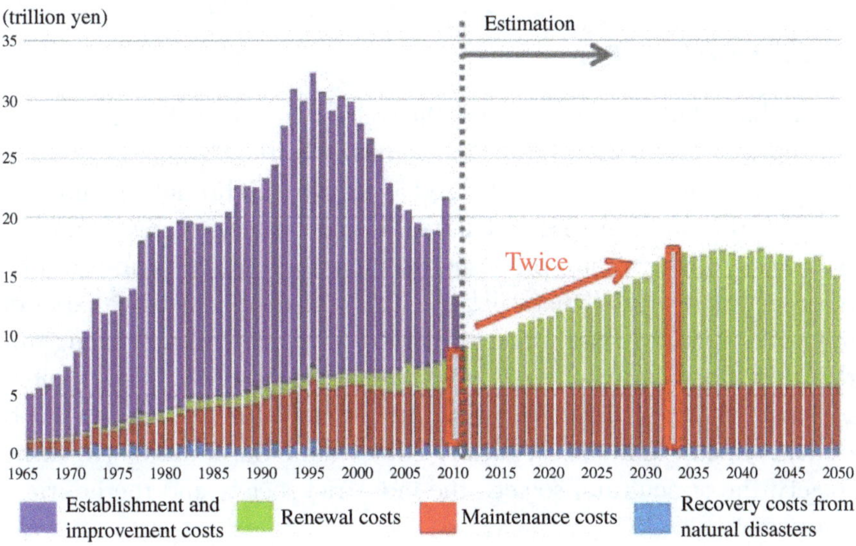

Figure 9.4. Expected increase in renewal and maintenance costs for infrastructure.

Source: From https://www.mlit.go.jp/common/000135837.pdf.

These challenges cannot be solved without an increase in productivity. For the declining birth rate and aging population, the utilization of women and elderly persons, along with robots, can enhance productivity. For the rise in labor-intensive maintenance costs for urban infrastructure, the introduction of new equipment which automatically surveys and detects problems with roads and water and sewage systems might increase productivity. Along with these policy measures, the use of ICT is essential for an increase in productivity to respond to such challenges. The use of ICT to overcome various societal challenges is also mentioned in the conceptual framework of Fernandez-Anez *et al.* [2018] in subcategory 6.1, "Enhancing the adaptation of society, governance and economy to transformation through ICT" within the category of "Technological innovations". This indicates that the conceptual framework by Fernandez-Anez *et al.* [2018] totally covered the current challenges faced by Japanese smart cities, indicating they are global trends. Therefore, this conceptual framework is evaluated as highly valid for smart city policies.

9.3.2 *Society 5.0*

The Cabinet Office first presented the concept of Society 5.0 in the Fifth Science and Technology Basic Plan approved by the Cabinet in 2016. In this plan, Society 5.0 is characterized as follows—"Through an initiative merging the physical space (real world) and cyberspace by leveraging ICT to its fullest, we are proposing an ideal form of our future society: A 'super-smart society' that will bring wealth to the people. The series of initiatives geared toward realizing this ideal society are now being further developed and intensively promoted as 'Society 5.0'" [Government of Japan, 2016, p. 13]. The most important feature of Society 5.0 is human wealth, and such a society comes after the hunting–gathering society, the agricultural society, the industrial society and the information society.

Society 5.0 is similar to the notions representing the fourth industrial revolution based on ICTs, such as "Industry 4.0" in Germany, "Advanced Manufacturing Partnership" in the United States, and

"Made in China 2025" in China. Whereas these concepts chiefly address industry itself, Society 5.0 represents society as a whole.

9.3.3 *Smart city policies after Society 5.0*

A larger difference between the first and second generations of smart city policies after Society 5.0 is the presence or absence of smart city platforms. In the first generation of smart city policies, namely the next-generation energy and social system demonstration project and the successor, the smart community project, both information and data on energy, transportation and other public services were shared and utilized through ICT systems within an individual public service sector. However, such information and data were not used across these service sectors. To overwhelm such shortages pertaining to the first generation of smart city policies, the smart city policies after Society 5.0 include the establishment of smart city platforms that enable us not only to obtain cross-sectoral data and information sharing, but also utilize Big Data derived from various data sources, including citizens. Examples of such platform-oriented projects are the promotion project for data utilization-driven smart cities by the Ministry of Internal Affairs and Communications in 2017 and the smart city model projects by the Ministry of Land, Infrastructure, Transport and Tourism in 2019. Both are briefly mentioned as follows.

9.3.3.1 *The promotion project for data utilization driven smart cities*

To solve various problems facing cities and regions and to revitalize local communities, this project subsidizes a portion of the expenses for the initial and continuous investments on cross-sectoral smart city development by local governments. In 2017, six cities—Sapporo, Aizu-wakamatsu, Saitama, Yokohama, Kakogawa, and Takamatsu cities—were designated for this project.

Regarding data sharing and utilization, many platforms for smart cities have been created. In Japan, the FIWARE is often used as a

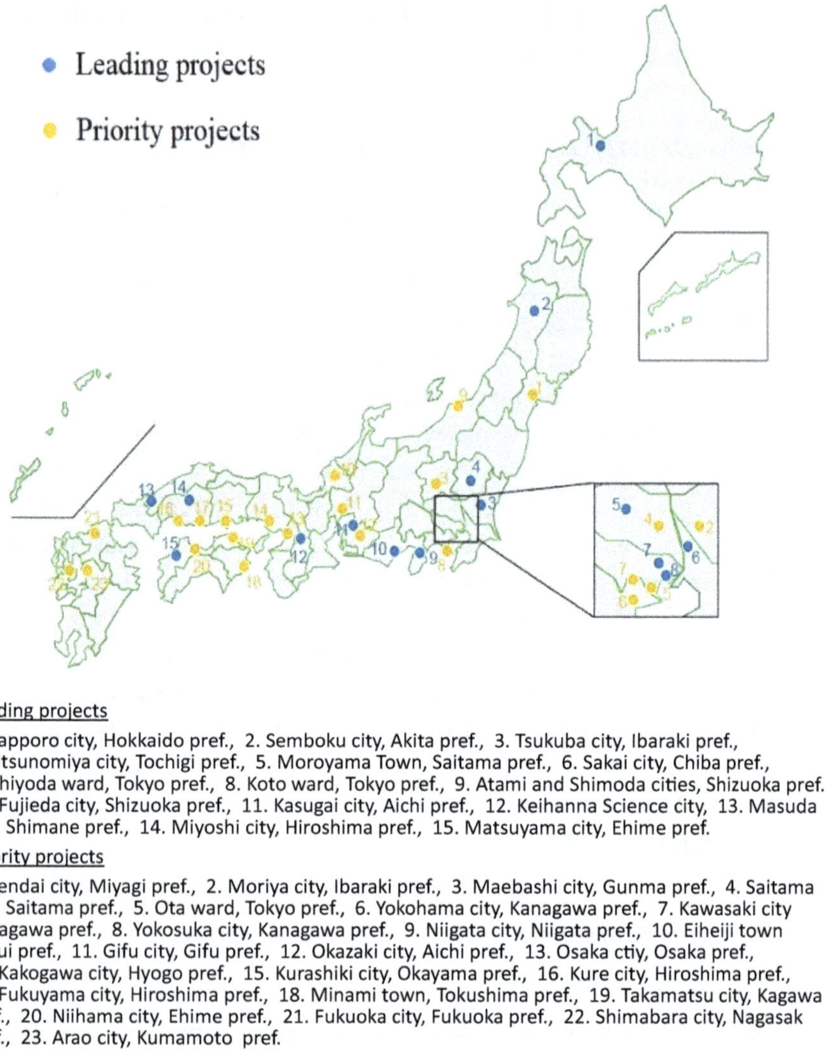

Figure 9.5. Designated cities of the smart city model project.
Source: From https://www.mlit.go.jp/common/001291681.pdf.

platform. Takamatsu city is the front-runner of the application of FIWARE-based platforms. This city has given a test bed for the FIWARE since 2017 in two public service sectors of tourism promotion and disaster management (Figure 9.6). In the Smart City Model

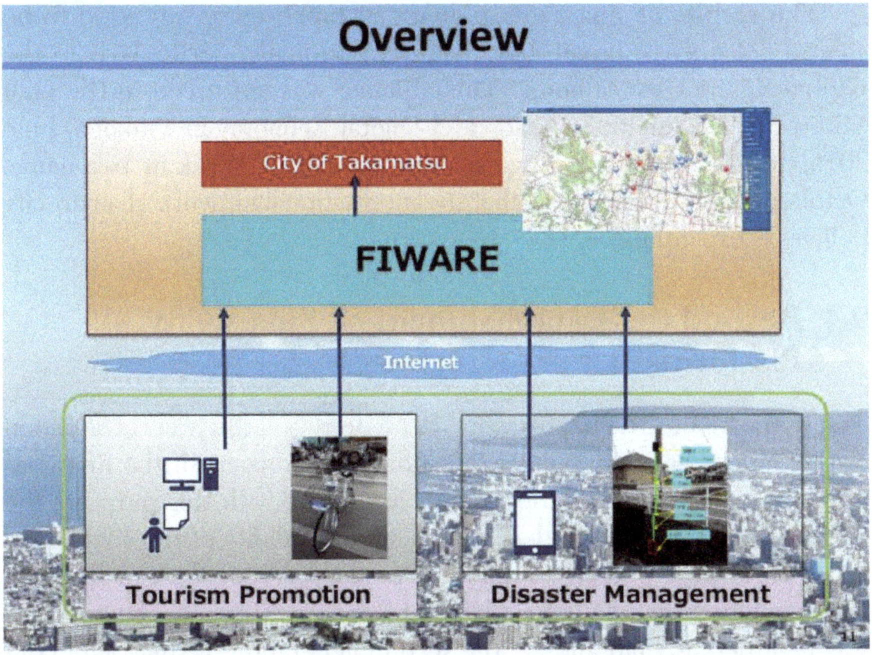

Figure 9.6. The utilization of FIWARE in the smart city Takamatsu project.

Projects, however, none of the 38 designated cities applied the FIWARE platform for environmental sustainability.

9.3.3.2 *The smart city model project*

Using new technology and public–private data, this project aims to implement solutions for sustainable and cross-sectoral urban development. Two types of projects, which are leading and priority projects, have been implemented since 2019. In the leading project, 15 cities were selected, while 23 were chosen in the priority projects (Figure 9.5). The leading projects are financially supported to back up both the leading and priority projects. In the leading projects, new programs in a project are launched, and program results and bottlenecks are analyzed. In addition, project results are shared with other projects. The priority projects are given national implementation support so that experts are dispatched and planning is also supported.

The results of smart city policies in Japan were intended to be shared not only in Japan, but also internationally, particularly in the Global Smart City Alliance. This alliance was approved in the G20 Ministerial Statement on Trade and Digital Economy in Tsukuba, June 2019, and established during the Asia Smart City Week in Yokohama, October 2019. This suggests that the international network of smart city policies is expanding.

9.4 Beyond the Establishment of Smart City Platforms

The purposes of this chapter were, firstly to reveal the present situation of the latest smart city policies, including the impacts of the first generation on these state-of-the-art policies. Along with this purpose, the present chapter also intended to evaluate the validity of the conceptual framework presented by Fernandez-Anez *et al.* [2018]. As a result, it was concluded that (1) the conceptual framework of smart city policies was regarded as valid and (2) the second generation of smart city policies after Society 5.0 was characterized by the establishment of smart city platforms. For the findings on the conceptual framework for smart city policies, however, it seems that further studies are necessary to confirm its validity in other countries.

Although the establishment of smart city platforms was identified to enhance data sharing among different policy domains about the impacts of smart city policies in Japan, sector-based projects are still observed among the second generation of smart city policies, just like in other countries [Mattoni *et al.*, 2015]. As installed in Fernandez-Anez *et al.* [2018]'s conceptual framework, governance is the key issue to bridge such sector-based smart city projects/policies [Albino *et al.*, 2015; Meijer and Bolívar, 2016]. Further studies are required to reveal how the smart city platform is utilized to foster governance in the second generation of smart city policies in Japan. In addition, the present chapter was dedicated to qualitative evaluation of the conceptual framework presented by Fernandez-Anez *et al.* [2018] for smart city policies. Further quantitative studies are necessary to confirm the validity of this framework, especially for the governance bridging the other five elements of Fernandez-Anez *et al.* [2018]'s conceptual framework using actual data.

References

Alawadhi, S. *et al.* (2012). Building understanding of smart city initiatives, in: Scholl, H.J., Janssen, M., Wimmer, M.A., Moe, C.E., and Flak, L.S. (eds.), *Electronic Government. EGOV 2012. Lecture Notes in Computer Science*, vol. 7443, Berlin, Heidelberg: Springer, https://doi.org/10.1007/978-3-642-33489-4_4.

Albino, V., Berardi, U., and Dangelico, R.M. (2015). Smart cities: Definitions, dimensions, performance, and initiatives, *Journal of Urban Technology*, *22*, 3–21.

Angelidou, M. (2014). Smart city policies: A spatial approach, *Cities*, *41*, 3–11.

Bower, L.L. (2019). The development impact of smart city programs: A case study: Istanbul's smarter city initiative, *Proceedings of the International Symposium on Technology and Society*, 20–25.

Caragliu, A., and Del Bo, C.F. (2019). Smart innovative cities: The impact of Smart City policies on urban innovation, *Technological Forecasting and Social Change*, *142*, 373–383.

Carayannis, E.G., and Campbell, D.F.J. (2009). 'Mode 3' and 'Quadruple Helix': Toward a 21st century fractal innovation ecosystem, *International Journal of Technology Management*, *46*, 201–234.

Castelnovo, W., Misuraca, G., and Savoldelli, A. (2016). Smart cities governance: The need for a holistic approach to assessing urban participatory policy making, *Social Science Computer Review*, *34*, 724–739.

Deakin, M. (2014). Smart cities: The state-of-the-art and governance challenge, *Triple Helix*, *1*, 1–16. DOI: 10.1186/s40604-014-0007-9.

Etzkowitz, H., and Zhou, C. (2006). Triple Helix twins: Innovation and sustainability, *Science and Public Policy*, *33*, 77–83.

European Parliament. (2014). *Mapping Smart Cities in the EU*, European Parliament.

Fernandez-Anez, V., Fernández-Güell, J.M., and Giffinger, R. (2018). Smart City implementation and discourses: An integrated conceptual model. The case of Vienna, *Cities*, *78*, 4–16.

Fernández-Güell, J.M., Collado-Lara, M., Guzmán-Araña, S., and Fernández-Añez, V. (2016). Incorporating a systemic and foresight approach into smart city initiatives: The case of Spanish cities, *Journal of Urban Technology*, *23*(3), 43–67.

Giffinger, R., Fertner, C., Kramar, H., Kalasek, R., Pichler-Milanovic, N., and Meijers, E. (2007). *Smart cities—Ranking of European medium-sized cities*, Centre of Regional Science, Vienna University of Technology.

Government of Japan. (2016). *The 5th Science and Technology Basic Plan*, Government of Japan.

Japan Smart Community Alliance. (2015). *SMART COMMUNITY—Japan's Experience*, Japan Smart Community Alliance.

Komiyama, Y. (2012). *Smart communities, as a Presentation of US–Japan Renewable Energy Policy Business Roundtable 'New Opportunities for Renewable Energy in Japan & the United States'*, held in Tokyo, Japan, December 3, 2012. URL: http://www.enecho.meti.go.jp/notice/event/036/pdf/event036013.pdf.

Kourtit, K., Nijkamp, P., and Steenbruggen, J. (2017). The significance of digital data systems for smart city policy, *Socio-Economic Planning Sciences*, *58*, 13–21.

Lombardi, P., Giordano, S., Caragliu, A., Del Bo, C., Deakin, M., Nijkamp, P., Kourtit, K., and Farouh, H., (2011). An advanced triple-helix network model for smart cities performance, *Green and Ecological Technologies for Urban Planning*, 59–73.

Marsal-Llacuna, M.L., Colomer-Llinàs, J., and Meléndez-Frigola, J. (2015). Lessons in urban monitoring taken from sustainable and livable cities to better address the Smart Cities initiative, *Technological Forecasting and Social Change*, *90*, 611–622.

Mattoni, B., Gugliermetti, F., and Bisegna, F. (2015). A multilevel method to assess and design the renovation and integration of smart cities, *Sustainable Cities and Society*, *15*, 105–119.

Meijer, A., and Bolívar, M.P.R. (2016). Governing the smart city: A review of the literature on smart urban governance, *International Review of Administrative Sciences*, *82*, 392–408.

Monzon, A. (2015). Smart cities concept and challenges: Bases for the assessment of smart city projects, *Communications in Computer and Information Science*, *579*, 17–31.

Promotion Council for the "FutureCity" Initiative. (2016). *"FutureCity" Initiative—Urban Development for the Future, Get it Rolling Regional Wisdom, Power, and Resources Will Strengthen Cities (2016 Edition), Promotion Council for the "FutureCity" Initiative*.

Viale Pereira, G., Cunha, M.A., Lampoltshammer, T.J., Parycek, P., and Testa, M.G. (2017). Increasing collaboration and participation in smart city governance: A cross-case analysis of smart city initiatives, *Information Technology for Development*, *23*, 526–553.

Yigitcanlar, T., and Kamruzzaman, M. (2018). Does smart city policy lead to sustainability of cities? *Land Use Policy*, *73*, 49–58.

© 2023 World Scientific Publishing Company
https://doi.org/10.1142/9789811257186_0010

CHAPTER 10

The Bright and Dark Sides of Open Innovation in Public Organizations: Evidence from Smart City Projects

Stefano Bresciani,[*] Alberto Ferraris,[†] Gabriele Santoro[‡,§]

[*]*Department of Management, University of Turin, Turin, Italy*
[†]*Department of Management, University of Turin, Italy and Ural Federal University, Ekaterinburg, Russia*
[‡]*Department of Management, University of Turin, Turin, Italy*
[§]*Corresponding author, Email: gabriele.santoro@unito.it*

10.1 Introduction

The development of efficient and effective public services in cities is one of the most important drivers for improving the quality of life of citizens and the overall society [Ferraris *et al.*, 2017; Paskaleva, 2011]. Due to steady and sudden technological and social changes [Phillips, 2011], improving the living conditions of citizens has become a priority for many mayors and local administrators who are under increasing pressure from different stakeholders. In this context, more people will be moving into the cities in the future, requiring new approaches to solve social and economic issues [Dekker and van Kempen, 2004]. Therefore, increasing globalization and technological progress [De Jong *et al.*, 2015] are also causing city managers to face more and more complex challenges. Public organizations should develop solutions and services that can meet stakeholders' demands, exploiting new approaches and methodologies. Hence, both public and private organizations may benefit from establishing open innovation practices and collaborative approaches [Oh *et al.*, 2016; West and Bogers, 2017]. The open innovation paradigm advocates that organizations should insource technologies and forms of knowledge/ideas to foster innovation [West and Bogers, 2014]. However, looking at the reference literature, most of the studies on open innovation so far focused on private organizations [Bogers *et al.*, 2018; Santoro *et al.*, 2017; Stanko *et al.*, 2017], thereby overlooking the context of public organizations [Santos and Mendonça, 2017], even though many public managers indicate that they recognize the potential of having an open approach.

By establishing open innovation practices, city managers can co-design new solutions with multiple actors, developing new services, reducing the entrepreneurial risk and exploiting resources that they lack [Nam and Pardo, 2011]. Quality and efficacy are improved through more targeted and better-tailored services, often implemented through partnerships with stakeholders. From a theoretical standpoint, the reasons behind the decision to share public projects and collaborate in the delivery of public value can be explained by the resource-based view (RBV) [Wernerfelt, 1984]. Accordingly, public organizations often lack of resources; sometimes, they also lack the capacity to identify citizens' and stakeholders' needs [Sandulli *et al.*, 2017]. As a consequence, open innovation allows cities to be more productive in developing innovative services by tapping into external resources [Betz *et al.*, 2016; Cohen *et al.*, 2016].

This model becomes vital in the smart city context, a system represented by modern communication infrastructures, i.e., information and communication technology (ICT), along with forms of human and social capital that allow increases in economic and sustainable development and the overall quality of life [Caragliu *et al.*, 2011], and an open approach to innovation [Ferraris *et al.*, 2018]. In the smart city context, citizens, public and private stakeholders, can engage with the ultimate goal of developing new solutions by connecting the physical, IT, social and business infrastructures and resources [Bresciani *et al.*, 2018; Hollands, 2008].

In line with these premises, this chapter deals with the emergence of the growing phenomenon of the "smart city" in which public organizations should have a leading role in the development of new and innovative "smart services" [e.g., Alves, 2013; Ferraris *et al.*, 2019; Mergel and Desouza, 2013]. Specifically, the present chapter builds on open innovation theory to explore how public organizations can improve the delivery of services through smart city projects. The main objectives of this chapter are (1) to highlight success factors in the co-development and co-delivery of new smart services with other cities' stakeholders, (2) to explore the dark side of open innovation experienced by public managers and (3) to merge these insights to present relevant and useful theoretical implications and practical takeaways.

In order to achieve these goals, this chapter explores the nature and dynamics of smart city projects through a qualitative research design involving interviews in order to understand the benefits and pitfalls of openness. This was done with different public governments in smart city initiatives through open questions and discussions with a panel of specialists in the field such as city managers, managers working in big and small companies, and academics. We used a mixed narrative and semi-structured interview track to investigate the best practices of and barriers to open innovation for public organizations with regard to established open innovation projects that had succeeded or failed.

The chapter contributes to theory and practice through three main guidelines. First, the chapter explores the open innovation phenomenon in public organizations, since almost all studies have focused on private organizations. Second, the study advances our knowledge by highlighting the barriers to open innovation, an under-investigated topic for both private and public organizations. Third, in line with the previous point, the study suggests key takeaways for managers and policymakers to overcome these barriers and to exploit the full potential of open innovation in the smart city context.

The next section reviews the literature on smart cities and open innovation, highlighting the role of specific actors involved in smart city projects, such as multinational companies, universities, and public organizations. The third section presents the research design involving interviews. The fourth section discusses the research findings regarding the success factors for and barriers to smart city projects. Finally, the last section highlights theoretical implications and key takeaways for managers and policymakers.

10.2 Literature Review

The smart city is a coherent urban system developed and managed by city governments seeking to plan and align in the long term the management of the various cities' infrastructural assets and municipal services with the objective of improving the quality of life of citizens [Dustdar *et al.*, 2017]. This system provides better opportunities for urban development by exploiting technological advances to accelerate the development of services [Appio *et al.*, 2019]. In fact, the advance of ICTs has offered a huge chance to modernize urban governance systems, to promote the development of smart services, and to tackle grand challenges within cities [Mora *et al.*, 2017, 2019; Nam and Pardo, 2011]. Although there is no doubt among academics and experts about these key aspects of smart cities, the reality is that many smart cities try to maximize efficiency while others focus more on creativity and experimentation (for example, through research parks, laboratories and ambitious projects carried out with multinational companies) to explore highly innovative solutions. As a result, many smart cities are experiencing growing tensions over the need to maximize efficiencies and reduce operating costs, while at the same time investing time and resources in innovative projects to solve bigger problems.

In this regard, the literature seems to suggest that smart cities with ambitious targets related to radical and disruptive innovation should focus on open innovation strategies involving different actors [e.g., Ferraris *et al.*, 2018]. Hence, open innovation is a paradigm that suggests that at the level of the individual organization, the flow of incoming and outgoing knowledge should be encouraged to support innovation processes [Chesbrough *et al.*, 2006]. From the RBV perspective, this paradigm indicates that no organization has all the resources in-house to innovate and therefore it is necessary to acquire externally those resources that they do not have internally. On the other hand, this paradigm suggests that unused or underused resources should be transferred externally for strategic or economic reasons (e.g., increasing turnover).

In the smart city context, each actor or partner has distinctive skills and technologies that contribute to the diversity needed to develop, nurture and test new solutions [Schaffers *et al.*, 2011; Scuotto *et al.*, 2016]. For example, it has been highlighted that universities act as knowledge intermediaries, knowledge gatekeepers, knowledge providers and knowledge evaluators [Ardito *et al.*, 2019]. In addition, the key role played by multinational companies in the smart city context has been found to be vital for providing unique tangible and intangible assets [Bresciani *et al.*, 2018]. In fact, multinational companies are the only ones that can bring technical knowledge and advanced technologies that can be used to test exploitative and explorative innovations in smart city contexts. Another study highlighted the importance of firms having interconnected explorative and exploitative human resource management systems to support social integration and knowledge management between internal and external employees in smart city projects [Ferraris *et al.*, 2019]. Moreover, partner selection in smart city projects has received considerable attention. In this sense, Sandulli *et al.* [2017] suggest that companies can improve the success of their smart city projects if they assess three main aspects of partner selection—(1) partner complementarity, (2) commitment and (3) compatibility.

As pointed out in the literature, most of the studies on smart cities have focused more on the technological aspects of the phenomenon and on the private side of the smart city partnerships (multinational companies, start-ups, etc.). As a result, part of the literature seems to pay less attention to the role of the public organizations, which should take the lead in smart city initiatives because they understand the needs of citizens and know the budget available for investing in smart city initiatives. To this end, Meijer and Bolívar [2016] stress the importance of developing a new smart city governance based on crafting new forms of human collaboration to achieve new outcomes and more open governance processes. The authors suggest that the development of solid and effective smart city projects should go through processes of institutional change and new forms of policy in favor of openness, experimentation and entrepreneurial risk in cities. This is in line with some previous studies indicating that public organizations are forced to engage with citizens to improve services [Cunningham and Kempling, 2009], and with scientific- and market-based partners to acquire unique resources and competencies [Appio *et al.*, 2019]. Despite this, public organizations often fail to adopt an entrepreneurial approach and to implement open strategies to make cities smart. This is mainly due to three factors [Ferraris *et al.*, 2020]—(1) a slow and cumbersome decision-making process [Rangan *et al.*, 2006; Rufin and Rivera-Santos, 2012], (2) the commitment to public services and therefore rigidity in pursuing certain objectives, which are often just related to cost reduction [Damanpour and Schneider, 2009; Yoon *et al.*, 2018] and (3) the possession of weak absorptive and innovation capabilities [Demircioglu and Audretsch, 2017]. Overall, these factors make public organizations unable to address smart growth paths linked to the establishment of collaborative strategies and risk-oriented approaches. As a consequence, it is reasonable to assume that there are barriers to the development of open innovation strategies in smart cities. However, there are some strategies and enabling factors that allow the overcoming these barriers.

10.3 Methods

The chapter builds on a qualitative research approach to achieve the mentioned objectives. The qualitative approach has been supported by semi-structured interviews performed over the phone and face-to-face [Alvesson *et al.*, 2008; Saunders *et al.*, 2009]. This methodology allowed us to explore facets regarding the bright and dark sides of open innovation in public organizations. As such, this can be considered an appropriate approach due to the limited knowledge on this context of analysis regarding public organizations and smart cities [Bogers *et al.*, 2018].

In addition, this research design allows exploring aspects and questions that are unexplored [Yin, 2013].

The research design consisted of a deep analysis of different public governments in smart cities' initiatives through open questions or discussions with a panel of specialists in the field. Due to the fact that smart city projects typically involve multi-stakeholders [Ferraris *et al.*, 2018], we decided to focus our attention on different kinds of experts (e.g., academics, public managers, firm's managers) who have experience in these projects and to integrate diverse perspectives regarding this phenomenon. The main focus of the interviews was on comprehending the role of open innovation in public organizations that develop smart city projects. More specifically, the research design involved semi-structured interviews [Cachia and Millward, 2011] to investigate the barriers to and success factors for open innovation in public organizations with regard to established open innovation projects that had succeeded or just failed. Semi-structured interviews have been chosen in order to let the participants shed light on other related aspects perhaps neglected by the interview protocol. We collected data from 20 interviews lasting two hours on average. Five of them were developed with smart city managers of multinational technology firms, while only three were from small- and medium-sized enterprises. Six of them were leading academics who co-developed projects with public municipalities and six of them were public managers.

Therefore, in order to have a broad and rational view of the barriers and enabling factors, we obtained insights from interviews done with both private and public actors. The projects can be considered as open innovation because they consist of several actors contributing different tangible and intangible resources. Some examples of the projects include the development of Big Data analytics on public-service monitoring; smart roads and drainage infrastructure; electrical and thermal energy control systems; subway control room; smart transportation systems; waste management and parking systems and intelligent street lighting. For each of these projects, the innovation approach of governments was debated from different angles and perspectives [Bresciani *et al.*, 2018]. These projects took place in several cities such as London (UK), Madrid (Spain), Milan (Italy), Ekaterinburg (Russia).

The authors conducted internal coding of information and both within-case and cross-case analyses of the interview transcripts to identify recurrent responses and themes [Miles and Huberman, 1994]. The outcome of this process is the identification of enabling factors and barriers to establishing open innovation by public organizations, in the smart city context.

10.4 The Success Factors in Smart City Projects

Overall, the smart city projects analyzed in this research were aimed at developing and introducing new services to improve the quality of life of citizens and communities in specific regions. Some of the projects were successfully developed and launched while others were stopped for several reasons. The interviews underlined several success factors, as follows.

10.4.1 *Open culture approach among city managers and citizens*

A mindset incorporating openness and collaboration is considered essential for developing new solutions to social problems and critical issues. The process of developing open strategies is often the starting point for the removal of barriers that might exist. The involvement of all relevant stakeholders within the organization is key to creating a culture of buy-in that goes beyond the creation of the Smart City Strategy. Ongoing coordination and sharing of challenges and successes as part of smart city governance is key to long-term success. Moreover, the self-commitment of city managers towards common causes improves the likelihood of developing open projects for social solutions. As a result, the culture of decision makers within public organizations has proven to be a trigger for smart city initiatives, insofar as this culture often determines the development of some riskier projects. This positive attitude towards "doing" rather than "waiting" favors city development schemes.

10.4.2 *Political commitment*

Smart city experiments need political commitment and leadership as well as the active participation of key officials. According to our interviews, city mayors played a key part in mobilizing organizational units and resources to support smart city projects. From our evidence, public managers rarely support experiments and innovations that may harm the political strategies of decision makers or the interests of significant stakeholders. While many pilot projects usually do not need strong political support, scaling up to the exploitation stage requires more than just political commitment; it is necessary to demonstrate not only to political leaders but also high-ranking officials of the municipality and the most influential stakeholders that there will be significant impacts of the new technologies or services on city performance.

10.4.3 *Governance structure to favor open innovation project*

The governance model of smart city is vital for the management of successful interorganizational projects.

Moreover, the governance should also ensure visibility and make desirable the adhesion and support of each stakeholder within the city's ecosystem. Public governments that want to secure good leads on smart city projects need to work on transversal policies that through the use of technologies propose to significantly develop the city and its development dynamics to include improve accessible services. Such a governance model should be built with an open dialogue with citizens, associations, companies and universities, through teamwork and a multidisciplinary steering committee able to build synergies between the policies on "vertical" themes.

10.4.4 *The key role of universities*

Our results highlight that when universities (or university personnel) have been involved, these projects generate great benefits. This is because of advantages in the knowledge/technology creation and transfers from the university to the project due to the skills and capabilities of researchers and professors involved that public governments usually do not have. Universities and polytechnics have proven significantly able to provide project management skills, business plan activities, and scientific and technical knowledge. Moreover, universities also play a knowledge brokerage role among different smart city projects, cross-fertilizing the city's innovation ecosystem, avoiding different technologies or ideas being developed by closed and small teams. Finally, the universities play a great role in social and societal involvement and ecosystem facilitation or networking, making open innovation in smart cities real and not just the words of public administrators.

10.4.5 *Understanding the problems and needs*

It is necessary to find out about citizens' problems in order to understand their needs before structuring formal partnerships and open projects. This market-driven approach to smart city projects can really foster the development of initiatives in line with citizens' needs and expectations. Listening and speaking to people can unveil many problems, such as access to mobile phones for the homeless. Teams should be sent to other cities to learn from their peers. Drawing on existing practices to gain inspiration has been seen as an important step toward a more successful and open public organization. Therefore, the city must adopt an entrepreneurial approach geared towards satisfying its citizens and their needs. This customer-centric approach must be equipped with tools and technologies that put data at the center of the decision-making process. City managers and employees should take advantage of surveys, interviews and data analytics to understand what the citizens' needs are, which of them are the most pressing and how to solve them.

10.4.6 *Lessons learned*

Connected to the previous point, it is important to spend some time learning from what has been done and what has worked (or not) for other cities. This benchmarking approach allows learning from the best practices of similar cities. Here, universities may help in sourcing for projects outside of a city and in connecting with other cities' municipalities. Another technique is to become actively involved in organizations that share knowledge, provide guidance and connect people to lessons learned from other regions. In such cases, collaborating with private firms that have been successfully involved in other cities' projects helps in establishing innovation processes. These firms have the chance to scale open innovation projects in different cities thus achieving economies of scale and scope. However, each city has its own identity, objectives, priorities and morphological, economic and social characteristics.

Therefore, cities should learn from other similar cities and adopt similar open innovation strategies.

10.4.7 *Going beyond consultancy activity in order to foster radical innovation*

Developing lasting and formal partnerships can be vital for going beyond mere incremental innovation or adjustment, in turn promoting radical and disruptive solutions through leveraging different resources and backgrounds of various actors involved in smart city projects. As the literature review pointed out, some cities are smart because they seek to maximize the efficiency of processes and activities. Others can be considered creative and entrepreneurial cities because they are constantly looking for radical solutions. The latter have become creative through collaboration with different stakeholders such as universities, multinational companies and startups, to develop radical innovations.

10.4.8 *Stakeholder benefits*

Assessing and communicating clear benefits attributed to each stakeholder is important for increasing the legitimacy of actions and improving the impact of projects. Projects that are developed based on real needs are more likely to have measurable outcomes and will likely create and disseminate value. This value can be assessed prior to the initiation of a project and transparently reported to stakeholders. As pointed out previously, cities should become customer-centric. However, the interviews underlined that it is necessary to analyze and communicate the benefits of smart city initiatives for different stakeholders, not just citizens. This helps to obtain the consent and participation of different stakeholders.

10.4.9 *Using performance indicators and a strategic approach*

While many of the specific project outcomes can be quantifiable, expected results may be less measurable, especially if there are many partners involved. However, it is important to use a number of Key Performance Indicators (KPIs) as reporting mechanisms over time for both internal operations as well as the outward message to citizens and partners. Tracking and reporting these in an easily understood format, perhaps through dashboards, is an important element of the measurement of open projects. In addition, the use of these KPIs must be consistent with the strategic orientation of the public organization/city, i.e., consistent with the vision and mission of the city.

10.4.10 *Alignment with regional and national public entities*

A Smart City Strategy should reinforce linkages and collaboration at a macro level of analysis (national or regional). Businesses and residents, including travelers, do not see city borders and expect continuity of services. At the same time, a successful strategy is one that addresses and reflects the needs and vision of the local community, including being customer-centric, focusing on enhancing the quality of life and providing a competitive advantage for businesses attracting positive growth. A starting point for a higher level of alignment could be developing common strategies with other regional and national entities.

10.5 The Dark Side of Smart City Projects

Open innovation strategies can appear as very effective models for innovating in the context of smart cities. However, these strategies require specific capabilities, culture, and approaches. In fact, organizations, both public and private, often fail to implement open innovation strategies. The main barriers to open innovation projects in smart cities are listed and discussed as follows.

10.5.1 *Lack of engagement needed to take responsibility*

Many cities, and specifically their managers, lack the engagement needed to take responsibility. This is because open innovation projects show tangible results only in the long run, with increasing costs in the short run. City managers are often under pressure by public opinion, which leads them to choose low-risk projects instead of ambitious ones. They thus prefer, in the short run, to limit risks by reducing costs driven by radical innovation projects. This means that, often time, public organizations focus on maximizing efficiency rather than stimulating creativity and technological development to make the city truly smart.

Moreover, several interviews suggested how ecosystems often rely on voluntary projects launched by and the personal dedication of employees, knowledge workers, city managers and public organizations' managers, highlighting the fragility of existing models of governance in sustaining innovation processes. In other words, the city becomes smart when the individuals who run it take responsibility for proposing ambitious projects, involving strategic partners.

10.5.2 *Slow speed of execution of processes, rules, and practices*

Linked to the previous point, public organizations are driven to create stability rather than change, so this could stop innovation processes again, especially radical and disruptive innovations. Open innovation requires change, new mindsets and culture, skills and experience, especially when open projects involve large companies that aim to increase profits. In addition, as highlighted by the interviews, public actors in the smart city need to be able to act quickly behind the pace of private organizations, which have a clear timeline and want tangible results in a tight timeframe.

10.5.3 *Lack of agile project management tools and skills*

Open innovation requires high-quality skills and managerial tools to identify needed resources, proper partners, assess business models and guide the project in its phases. Often, the public organization does not have these elements. Moreover, it is often hard for public organizations to embrace customer-centric approaches to innovation, which is a vital starting point for smart city initiatives and innovation projects.

10.5.4 *Lack of resources, which increases organizational inertia and path-dependency effects*

The lack of resources has been seen as a limitation in developing joint projects but sometimes an incentive to become involved in open innovation projects. However, resource limitation encourages openness mostly when the organization possesses search capabilities aimed at looking the right partners and managing relationships accordingly. In other words, public management must be able to understand the needs of the citizens, to identify the problems of the territory, to identify what resources are needed to acquire and which partners to involve. This, again, passes through the ability of individuals, often city managers, to identify which companies can really bring value to smart city projects related to the problems of the territory and citizens.

10.5.5 *Public procurement rules*

Most of the projects in traditional cities are based on old models of financing which consist of annual budget cycles. This financial management decreases flexibility, especially in the case of unexpected business and technological opportunities that may be encountered by city managers. There is thus a need for new participatory and flexible models of project financing, such as joint investments and shared ownership, providing skills rather than mere subsidies. In addition, there is a need to include agile financing modes to ensure that unexpected opportunities can be financed and developed.

10.5.6 *Risk adversity*

It is well known that security and safety goals are the priorities of public systems. While errors in the private sector are frequent and often useful for improving learning and adaptation, errors in the public sphere are not accepted because politicians risk their reputations and even their jobs if they lose the next political election. This causes a detrimental effect on innovation. It is well known that politicians often prefer short-term strategies because they have a short mandate to focus on ambitious long-term projects.

Table 10.1. Key findings.

Success factors	Barriers
Open culture approach among city managers and citizens	Lack of engagement needed to take responsibility
Political commitment	Slow speed of execution of processes, rules and practices
Governance structure to favor open innovation project The key role of universities	Lack of agile project management tools and skills
Understanding the problems and needs	Lack of resources, which increases organizational inertia and path-dependency effects
Lessons learned	Public procurement rules
Going beyond consultancy activity in order to foster radical innovation Stakeholders' benefits Using performance indicators and a strategic approach Alignment with regional and national public entities	Risk adversity

10.6 Discussion, Implications and Conclusions

This chapter aimed at shedding light on a particular and relevant topic in the current era which faces new issues and social phenomena [e.g., Ferraris *et al.*, 2017; Mergel and Desouza, 2013]. More specifically, the research presented in this chapter aimed to (1) highlight success factors in the co-development and co-delivery of new smart services with other cities' stakeholders, (2) explore the barriers to open innovation experienced by public organizations and (3) merge these insights to present relevant and useful theoretical implications and practical takeaways (see Table 10.1). Through a qualitative methodology, we have found that success factors for implementing open innovation in public organizations are—an open culture approach to innovation; a strong political commitment; a proper governance structure to favor openness; the presence of universities within the projects; an openness toward listening to stakeholders and understanding the city's problems; an attitude toward learning from other projects; establishing formal and steady partnerships for radical innovations; assessing and communicating the impacts of projects for each stakeholder; using KPIs to assess the performance of projects and linking them to the strategic orientation; and aligning strategies with national and regional entities. By contrast, we have outlined several barriers to implementing open innovation in public organizations, such as—lack of engagement needed to take responsibility; a slow speed of execution of projects; a lack of project management tools and skills from the public side; a lack of resources; the limits of public procurement rules; and a high risk adversity that limits innovative projects.

These findings help us to draw some important theoretical insights and implications for practice. From a theoretical point of view, the chapter offers the following implications. First, the chapter adds to the literature on open innovation [Bresciani, 2017; Chesbrough *et al.*, 2006] by offering an overview of the characteristics and dynamics that favor and limit the adoption of open approaches in public organizations. Indeed, previous studies on open innovation mostly focused on private organizations [Bogers and Jensen, 2017; Santoro *et al.*, 2018]. Moreover, the study focuses on the so-called bright and dark sides of open innovation, since much of the previous literature has focused on the benefits of open innovation, such as better financial, economic and innovative performance [Laursen and Salter, 2006]. To this end, this study suggests which factors are favoring and holding back the development of open strategies. Open innovation is a managerial approach oriented toward the co-development of innovative solutions [Wang *et al.*, 2012], and innovative business models [Franceschelli *et al.*, 2018]. This managerial approach requires management, organization, objectives and alignment with partners and, therefore, can cause tensions and difficulties from an organizational point of view. Precisely for this reason it is necessary, from a theoretical standpoint, to shed light on which aspects limit this approach. Second, the chapter adds to the current body of knowledge on smart city research [Appio *et al.*, 2019; Nam and Pardo, 2011], exploring how open innovation strategies in public organizations may lead to changes and how they entail barriers and success factors.

Specifically, using a practical and inductive approach, the study explores the lights and shadows of collaborative projects in the context of smart cities—a context that forces the use of collaborative approaches to solve citizens' problems and improve the services offered [Ahvenniemi *et al.*, 2017; Albino *et al.*, 2015].

The study also offers clear and precise takeaways for different stakeholders in smart cities. The smart city and its development depend on the ability to involve and encourage a range of different actors. In particular, a smart city cannot exist without a model of dialogue among the various stakeholders that provide public services and offer their assets and resources. Public governments should be able to improve the quality of life of its citizens by using integrated technological solutions from a solid organizational culture and ensure individual rights, the care of common areas (including digital ones) and respect for public spaces. In a nutshell, the research findings of this study suggest the following:

(1) *Adopting a customer-centric approach to city management, listening to the opinions and understanding the needs of citizens*. This should be the starting point for the development of smart city projects useful for improving citizens' quality of life.
(2) *Adopting a strategic approach to project management*. This means developing explicit objectives to be achieved, analyzing the necessary resources before starting projects, and including internal and external stakeholders. The strategic approach should be in line with the vision, mission and identity of the city. Some successful projects have been developed with the process management approach that follows the following stages—problem identification, solution proposal, identification of resources needed, partners' identification and outcome.
(3) There is a need to increase the number and intensity of international partnerships to gain both legitimacy and good repute that can create a virtuous circle of investments, collaborations, innovative projects and stakeholders included, regardless of the short-term benefits of the partnership itself. These collaborative approaches could exploit practices, spaces and environments such as hackathons, incubators, accelerators, shared spaces and living labs. Open innovation

strategies should include collaboration with different actors such as universities, multinational companies and startups.

(4) It is necessary to foster an experimental, agile, and test-and-learn thinking approach to improve entrepreneurial processes within public systems and to accelerate decision-making processes. As pointed out in the interviews, mayors or city managers often choose short-term, low-risk projects in order to avoid not being re-elected at the end of their term. However, this approach holds back growth and the nurturing of a creative and entrepreneurial city. Mayors and city managers should therefore balance short-term projects with long-term projects, less risky projects with riskier projects. In any case, the city should support entrepreneurial approaches and risk culture.

A major limitation of the study concerns subjectivity in the responses and thus in the gathering of data. However, consistent with our interpretive and constructivist epistemological stance, we are aware that our research activity is a means of 'world making' [Brown *et al.*, 2008] that co-produces a particular version of reality together with our research participants.

In conclusion, open innovation in public organizations, and especially in the smart city context, is still in its infancy. While there are many studies addressing the differences in innovative activities between private and public organizations, studies on the differences in the specific context of open innovation activities are missing. In fact, while in this study we focus on just the public side, future studies could address both sides, highlighting differences. Moreover, our research suggests that open innovation poses challenges to public organizations and that there are several barriers to implementing it. Finally, future studies could focus on open innovation in public organizations that aim to develop social and sustainable solutions [Bogers *et al.*, 2020].

References

Ahvenniemi, H., Huovila, A., Pinto-Seppä, I., and Airaksinen, M. (2017). What are the differences between sustainable and smart cities? *Cities*, *60*, 234–245.

Albino, V., Berardi, U., and Dangelico, R.M. (2015). Smart cities: Definitions, dimensions, performance, and initiatives, *Journal of Urban Technology*, *22*(1), 3–21.

Alves, H. (2013). Co-creation and innovation in public services, *The Service Industries Journal*, *33*(7–8), 671–682.

Alvesson, M., Hardy, C., and Harley, B. (2008). Reflecting on reflexivity: Reflexive textual practices in organization and management theory, *Journal of Management Studies*, *45*(3), 480–501.

Appio, F.P., Lima, M., and Paroutis, S. (2019). Understanding Smart Cities: Innovation ecosystems, technological advancements, and societal challenges, *Technological Forecasting and Social Change*, *142*, 1–14.

Ardito, L., Ferraris, A., Petruzzelli, A.M., Bresciani, S., and Del Giudice, M. (2019). The role of universities in the knowledge management of smart city projects, *Technological Forecasting and Social Change*, *142*, 312–321.

Betz, F., Carayannis, E., Jetter, A., Min, W., Phillips, F., and Shin, D.W. (2016). Modeling an innovation intermediary system within a helix. *Journal of the Knowledge Economy*, *7*(2), 587–599.

Bogers, M., and Jensen, J.D. (2017). Open for business? An integrative framework and empirical assessment for business model innovation in the gastronomic sector, *British Food Journal*.

Bogers, M., Chesbrough, H., and Moedas, C. (2018). Open innovation: Research, practices, and policies, *California Management Review*, *60*(2), 5–16.

Bogers, M., Chesbrough, H., and Strand, R. (2020). Sustainable open innovation to address a grand challenge, *British Food Journal*.

Bresciani, S. (2017). Open, networked and dynamic innovation in the food and beverage industry, *British Food Journal*.

Bresciani, S., Ferraris, A., and Del Giudice, M. (2018). The management of organizational ambidexterity through alliances in a new context of analysis: Internet of Things (IoT) smart city projects, *Technological Forecasting and Social Change*, *136*, 331–338.

Brown, A.D., Stacey, P., and Nandhakumar, J. (2008). Making sense of sensemaking narratives, *Human Relations*, *61*(8), 1,035–1,062.

Cachia, M., and Millward, L. (2011). The telephone medium and semi-structured interviews: A complementary fit, *Qualitative Research in Organizations and Management: An International Journal*, *6*(3), 265–277.

Caragliu, A., Del Bo, C., and Nijkamp, P. (2011). Smart cities in Europe, *Journal of Urban Technology*, *18*(2), 65–82.

Chesbrough, H., Vanhaverbeke, W., and West, J. (eds.) (2006). *Open Innovation: Researching a New Paradigm*. Oxford University Press on Demand.

Cohen, B., Almirall, E., and Chesbrough, H. (2016). The city as a lab: Open innovation meets the collaborative economy, *California Management Review*, *59*(1), 5–13.

Cunningham, J.B., and Kempling, J.S. (2009). Implementing change in public sector organizations, *Management Decision*, *7*(2), 330–344.

Damanpour, F., and Schneider, M. (2009). Characteristics of innovation and innovation adoption in public organizations: Assessing the role of managers, *Journal of Public Administration Research and Theory*, *19*(3), 495–522.

De Jong, M., Joss, S., Schraven, D., Zhan, C., and Weijnen, M. (2015). Sustainable–smart–resilient–low carbon–eco–knowledge cities; making sense of a multitude of concepts promoting sustainable urbanization, *Journal of Cleaner Production*, *109*, 25–38.

Dekker, K., and van Kempen, R. (2004). Urban governance within the Big Cities Policy: Ideals and practice in Den Haag, the Netherlands, *Cities*, *21*(2), 109–117.

Demircioglu, M.A., and Audretsch, D.B. (2017). Conditions for innovation in public sector organizations, *Research Policy*, *46*(9), 1,681–1,691.

Dustdar, S., Nastić, S., and Šćekić, O. (2017). *Smart Cities. In The Internet of Things, People and Systems*. Springer.

Ferraris, A., Erhardt, N., and Bresciani, S. (2019). Ambidextrous work in smart city project alliances: Unpacking the role of human resource management systems, *The International Journal of Human Resource Management*, *30*(4), 680–701.

Ferraris, A., Santoro, G., Bresciani, S., and Carayannis, E.G. (2018). HR practices for explorative and exploitative alliances in smart cities: Evidences from smart city managers' perspective, *Management Decision*, *56*(6), 1,183–1,197.

Franceschelli, M.V., Santoro, G., and Candelo, E. (2018). Business model innovation for sustainability: A food start-up case study, *British Food Journal*.

Hollands, R.G. (2008). Will the real smart city please stand up? Intelligent, progressive or entrepreneurial? *City*, *12*(3), 303–320.

Laursen, K., and Salter, A. (2006). Open for innovation: The role of openness in explaining innovation performance among UK manufacturing firms, *Strategic Management Journal*, *27*(2), 131–150.

Meijer, A., and Bolívar, M.P.R. (2016). Governing the smart city: A review of the literature on smart urban governance, *International Review of Administrative Sciences*, *82*(2), 392–408.

Mergel, I., and Desouza, K.C. (2013). Implementing open innovation in the public sector: The case of Challenge.gov, *Public Administration Review*, *73*(6), 882–890.

Miles, M.B., and Huberman, A.M. (1994). *Qualitative Data Analysis: An Expanded Sourcebook*. Sage.

Mora, L., Bolici, R., and Deakin, M. (2017). The first two decades of smart-city research: A bibliometric analysis, *Journal of Urban Technology*, *24*(1), 3–27.

Mora, L., Deakin, M., and Reid, A. (2019). Strategic principles for smart city development: A multiple case study analysis of European best practices, *Technological Forecasting and Social Change*, *142*, 70–97.

Nam, T., and Pardo, T.A. (2011, September). Smart city as urban innovation: Focusing on management, policy, and context, in: *Proceedings of the 5th International Conference on Theory and Practice of Electronic Governance*, pp. 185–194. ACM.

Oh, D.S., Phillips, F., Park, S., and Lee, E. (2016). Innovation ecosystems: A critical examination, *Technovation*, *54*, 1–6.

Paskaleva, K.A. (2011). The smart city: A nexus for open innovation? *Intelligent Buildings International*, *3*(3), 153–171.

Phillips, F. (2011). The state of technological and social change: Impressions, *Technological Forecasting and Social Change*, *78*(6), 1,072–1,078.

Rangan, S., Samii, R., and Van Wassenhove, L.N. (2006). Constructive partnerships: When alliances between private firms and public actors can enable creative strategies, *Academy of Management Review*, *31*(3), 738–751.

Rufin, C., and Rivera-Santos, M. (2012). Between commonweal and competition: Understanding the governance of public–private partnerships, *Journal of Management*, *38*(5), 1,634–1,654.

Sandulli, F.D., Ferraris, A., and Bresciani, S. (2017). How to select the right public partner in smart city projects, *R&D Management*, *47*(4), 607–619.

Santoro, G., Vrontis, D., and Pastore, A. (2017). External knowledge sourcing and new product development, *British Food Journal*.

Santoro, G., Vrontis, D., Thrassou, A., and Dezi, L. (2018). The Internet of Things: Building a knowledge management system for open innovation and knowledge management capacity, *Technological Forecasting and Social Change*, *136*, 347–354.

Santos, A.B., and Mendonça, S. (2017). Open innovation in clusters: The Portuguese case, in: Monteiro, S., and Carayannis, E. (eds.), *The Quadruple Innovation Helix Nexus: A Smart Growth Model, Quantitative Empirical Validation and Operationalization for OECD Countries*, pp. 245–264. Palgrave MacMillan.

Saunders, M., Lewis, P., and Thornhill, A. (2009). *Research Methods for Business Students*. Pearson Education.

Schaffers, H., Komninos, N., Pallot, M., Trousse, B., Nilsson, M., and Oliveira, A. (2011, May). Smart cities and the future internet: Towards cooperation frameworks for open innovation, in: *The Future Internet Assembly*, pp. 431–446. Berlin, Heidelberg: Springer.

Scuotto, V., Ferraris, A., Bresciani, S., Al-Mashari, M., and Del Giudice, M. (2016). Internet of Things: Applications and challenges in smart cities. A case study of IBM smart city projects, *Business Process Management Journal*, 22(2). URL: https://doi.org/10.1108/BPMJ-05-2015-0074.

Stanko, M.A., Fisher, G.J., and Bogers, M. (2017). Under the wide umbrella of open innovation, *Journal of Product Innovation Management*, *34*(4), 543–558.

Wang, Y., Vanhaverbeke, W., and Roijakkers, N. (2012). Exploring the impact of open innovation on national systems of innovation—A theoretical analysis, *Technological Forecasting and Social Change*, *79*(3), 419–428.

Wernerfelt, B. (1984). A resource-based view of the firm, *Strategic Management Journal*, *5*(2), 171–180.

West, J., and Bogers, M. (2014). Leveraging external sources of innovation: A review of research on open innovation, *Journal of Product Innovation Management*, *31*(4), 814–831.

West, J., and Bogers, M. (2017). Open innovation: Current status and research opportunities, *Innovation*, *19*(1), 43–50.

Yin, R. K. (2013). Validity and generalization in future case study evaluations, *Evaluation*, *19*(3), 321–332.

Yoon, H.D., Kim, N., Buisson, B., and Phillips, F. (2018). A cross-national study of knowledge, government intervention, and innovative nascent entrepreneurship, *Journal of Business Research*, *84*, 243–252.

© 2023 World Scientific Publishing Company
https://doi.org/10.1142/9789811257186_0011

CHAPTER 11

Why Do Regional Innovation Systems Not Have Comprehensive Information Support?

Fred Young Phillips,[*,§] Benjamin Matheson,[†] Urusha Thapa,[†] Yun Jeong Choi,[‡] J. Byun[‡]

[*] *State University of New York, Stony Brook, USA*
[†] *University of New Mexico, Albuquerque, USA*
[‡] *KISTI Technology Commercialization Center, Daejeon, South Korea*
[§] *Corresponding author, Email: fred.phillips@stonybrook.edu*

11.1 Introduction: Regional Innovation Systems and their Information Requirements

A regional innovation system (RIS) consists of entrepreneurs, investors, educators, service providers, and a host of other enterprises and institutions. A growth-oriented RIS requires intensive interaction among these actors, for the purposes of encouraging new inventions and new ventures, supporting existing local companies, and attracting relocation and expansion of existing knowledge-based businesses. It stands to reason that these purposes are best served by making it easy for legacy actors, new actors, and prospective actors to learn of each other's existence and to profile each other. This is an ideal role for an information system (IS).

In investigating this situation for the Korea Institute for Science and Technology Information (KISTI), we found a surprising scarcity of comprehensive ISs serving RISs in the US and South Korea. Rather, we found that ISs supporting RISs are partial, fragmented, user-unfriendly, or less than comprehensive in their topic coverage.

In this chapter, we catalog the obstacles to comprehensive Information Infrastructure for Regional Innovation Systems (henceforth, "IIRIS"). We describe these both for grassroots RIS initiatives (the non-governmental initiatives more common in the US) and for government-led RIS initiatives (more common in Korea), and investigate the ways in which the obstacles to IIRIS differ between the two initiative types.

We distinguish an RIS from an "RIS initiative", the latter being a government or grassroots organization dedicated to growing the RIS. Often, the RIS initiative is an umbrella agency attempting to promote (or even coordinate) the efforts of a number of RIS support organizations. These agencies are the actors that may be expected to build and maintain information systems supporting the RIS.

Our mixed methodology relies on interviews with RIS initiative leaders and actors, examination of a variety of RIS web sites, and literature search. We draw as well on our own consulting experience [Oh *et al.*, 2016; Phillips 2006, 2008; Phillips *et al.*, 2014, 2019].[1] These lead to a top-level design of an "ideal" comprehensive IIRIS, detailed in the chapter, which we use as a benchmark to assess the quality of the examined RISs. Though evidence from other RISs is cited, our comparative analysis centers largely on Daejeon/Daeduk in Korea, and Portland, Oregon and Austin, Texas in the US.

In the course of this project, we found, contrary to our expectations, fragmented information sources in bottom-up RISs, and missing or largely unused portals in government-driven RISs. While the portals of a prominent Korean RIS (Daeduk) no longer feature bureaucratic language in unattractive formats (as is still the case with government-led IIRIS elsewhere in Asia), they are nonetheless rarely updated and seldom used. Western IIRISs tend to be fragmented, i.e., distributed over portals maintained by many non-profit organizations (NGOs). This is because all local NGOs compete for funding from governments and foundations, and thus want to differentiate themselves. Neither

[1] One of us (Phillips) played varied roles in the techno-economic development of Austin, Portland, San Diego, and Songdo (Korea).

arrangement, West or East, is conducive to an efficient exchange of information.

11.1.1 *Order of the presentation*

In what follows, we define RISs, distinguish grassroots RIS initiatives from top-down initiatives, and introduce the idea of RIS maturity. We then list the kinds of information support an RIS needs, drawing on the rather sparse literature, and display an ideal, "best practice" information architecture. We describe the information infrastructures of Austin, Portland, and Daeduk, comparing them to the ideal. Introducing highlights of our expert interviews, we categorize and summarize the reasons RISs do not appear to have adequate information support. Finally, we remark on connections between the topics of IIRIS and the "smart city", and offer research conclusions and action recommendations for RISs. Because the recommendations require cultural and regulatory reform, only the most flexible regions, having long-term orientation, will be able to heed them.

11.1.2 *Regional Innovation Systems defined*

An innovation system is made up of multiple groups, such as institutions, companies, non-government organizations (NGOs), laboratories, public opinion, and the triple-helix actors (government–industry–academia), all participating in mutual communication, creation, and learning. A regional innovation system (RIS) creates new knowledge and shapes innovation capacities in a region larger than a city, but usually smaller than a nation. A successful, synergistic [Leydesdorff *et al.*, 2019] regional innovation system excels at translating scientific discoveries into new technologies, products, and services. Such an RIS consists of one or more industry clusters [Porter, 2003] and is alternately known as a technopolis or an innovation ecosystem.

Precursors to this kind of success include [Niosi, 2010] human capital, cumulative expenditure on research and development (R&D), social capital [Phillips, 2006], patents and articles per capita, and openness to existing outside knowledge.

11.1.3 *Grassroots-driven RIS*

Grassroots-driven RISs are fostered by meso-level (neither family organizations nor government agencies) associations. They exemplify the region's social capital, i.e., the population's propensity to form voluntary, meso-level cooperative endeavors. The grassroots initiative members may see the number of knowledge-based industries growing locally, as happened in Austin, and organize to boost it farther. Or, they may see their home city lagging behind—a Maturity Level 1 area—and organize to "start from zero", a very difficult job.

Grassroots initiative-takers have grasped the concept of innovation, many having direct prior experience in policy, entrepreneurship, investment, or invention. The committed individuals search for creative ideas and influential support for innovation through grant funding, voluntary input, and mutual exchanges. These enthusiastic individuals engage with research institutions, governments, and private enterprises.

Completely lacking in authority, they are confident they can start a parade and that the government will then march alongside.

11.1.4 *Government-driven RIS*

In the government-driven RIS, government ministries and agencies play the chief role in building a robust science/technology/commercialization capacity for the region's economic success. Governments affect land planning, oversee universities, build research parks, and support innovation in strategic industries.

South Korea, one of the most innovative countries in the world, is a leading practitioner of government-driven RIS. The government of the country continues to support the technology field with a strategy to create jobs and to create a growing economy by increasing the pace of innovation. Representatives from dozens of countries travel to Korea yearly to observe the best government-driven RIS practices [Oh and Phillips, 2014].

11.1.5 *Technopolis maturity levels*

Regions begin their innovation journeys at varying levels of wealth, political and economic stability, social and intellectual capital, and cultural openness, and progress to higher levels of these at different rates. We prepared our interviewees for an RIS maturity measurement instrument assessing a region's Knowledge Base, Industry, Social Capital, Institutional Capital and Quality of Life, Relational Capital, and Access to Finance.[2]

11.2 The Literature

We were surprised to find next to no literature on information support for RISs. Almost alone in that empty field, Townsend, Pang, and Weddle [2009] present three interesting scenarios for the future of RISs:

> In the first scenario, an incremental evolution of the research parks model takes place in a world of rapid, but steady and predictable change. In the second scenario, entirely new networks of R&D space emerge in a "research cloud" that challenges current models to adapt, sometimes dramatically. The third scenario, the research park models is in rapid decline as R&D becomes highly virtualized and parks' legacy cost structure makes them obsolete for young firms.

The first scenario appears unrealistic; the second and third imply an increased role for IIRISs.

Tassey [2019] confirms that "regional growth policy [responds] to a set of ongoing trends in the rate and directions of technological change, global competition, and economic growth policy philosophies at the federal and state levels. A major characteristic of these policies is that they depend on access to a highly sophisticated technology infrastructure".

It is critical for there to be a high level of connection among all those who participate in the transfer of knowledge within a RIS [Niosi, 2010].

---

[2] General Informatics LLC. (2014). *Technopolis Maturity Assessment*. URL: www.generalinformatics.com/GI_Cluster_Maturity_Model_Evaluation_Chart.pdf.

For participants to successfully interact, there needs to be an information system that is suitable and affordable for both emerging and established RISs. Information infrastructure for regional innovation systems (IIRISs) should be practicable for RISs operating under any of the RIS regimes in the classification of Amable *et al.* [1997]:

- *Market*: USA, UK, Canada, Australia, New Zealand.
- *Social democratic*: The Nordic countries.
- *Continental Western European*: France, Germany, Italy, Spain.
- *Meso-corporatist*: Japan, South Korea.

In this chapter, however, we distinguish only the two RIS types—(1) government-driven RISs that fall under Meso-corporatist and (2) grassroots-driven RISs that fall under Market.

Building trust on the part of all those who participate in RISs, to the point where the actors are willing to depend on an IIRIS, can be a slow process [Belanche *et al.*, 2012]. Establishing a successful IIRIS requires face-to-face contact among those involved, in order to first build trust. Therefore, an IIRIS should share many characteristics with existing online collaboration systems[3] yet be highly attractive and useful, in order to engage participants. The IIRIS should focus on the local region but should also connect to other regions for the exchange of instructive success and failure stories.

A successful RIS promotes the translation of knowledge into new technologies through the support of an information infrastructure [Schot and Steinmeuller, 2018].

11.3 What Information Support does a Regional Innovation System (RIS) Need?

As the present authors have assessed client needs, it appears to us that an IIRIS that successfully advances the technology economy should support trust-building and intellectual property protection. It should have compelling content, much of it user-generated.

[3]The present chapter is also informed by many conversations with T. Schifter, CEO of TallyFox Social Technologies AG, makers of online collaboration systems.

It should be operated and actively moderated by an unbiased intermediary. Depending on local circumstances, this may be a university, a public library, a cluster initiative organization, a professional society, or a similar organization that is widely perceived as unbiased and free from conflicts of interest. Again, depending on local conditions, the local government may or may not be an appropriate IIRIS administrator. The moderator will curate content and periodically remind the actors of the benefits of using the system.

An IIRIS must be organized in such a way that it is not viewed by actors as just another website or mobile app to log on to every day, making their lives needlessly more complicated. The content must be compelling and appealing to all of those involved with RISs (i.e., politicians, civic and professional leaders, company executives, engineers, professors, entrepreneurs, and students).

An entrepreneur is on the road constantly in order to meet with investors, suppliers, (potential) customers, and service providers (i.e., lawyers and accountants, etc.) and to attend educational and professional networking events. Thus, the IIRIS must also be mobile-friendly.

Other prospective IIRIS features that were discussed with our interviewees include:

- Shared tools
- Virtual meetings and teleconferencing
- Access to market and technology forecasts
- Online library and educational materials (curated)
- Links to industry news
- STEM and entrepreneurship materials
- Job postings
- Event planning templates.

11.3.1 *An ideal IIRIS*

With the help of the interviewees, we devised a specification for an ideal IIRIS. This specification is clearly "overkill" for the less mature RISs. Yet, it provides a benchmark against which to measure the varied RIS's information infrastructures.

This full-featured IIRIS will contain 10 modules, each with several functions, as follows:

(1) *Guide Module*: This section is typically the first display seen by the user on an information system and is often referred to as the go-to window. The display has tabs that can be "clicked" on by the user as a way to redirect to information specific to research, local businesses, service providers (i.e., lawyers, accountants, IT services), etc. It is important for this section to be easily navigable, presenting no undue challenges for searchers.
(2) *Metrics/Evaluation Module*: This section will include reports on, e.g., educational progress, crime reduction, tax incentive program performance, number of companies formed, number of jobs created; in other words, all indicators of the region's progress (relative to its chosen role model RISs). It is important for this section to be up-to-date. It should rely on infographics to make it easy for the user to understand data reports.
(3) *Moderation/Social Engineering Module*: This module is the primary venue for the human moderator(s) to encourage wide participation in the IIRIS. Via this module, the moderator encourages users to contribute and access information. The moderator's messages through this module build trust and engagement with the user community, and suggest profitable connections among users and among potential users. The moderators' messages motivate the community to bend efforts toward RIS success. He/she ensures the IIRIS's data are in a format that member companies can use.
(4) *Collaboration Module*: This module allows users to direct-message other users, and encourages users to share different types of information. For example, this site typically includes options for chat rooms, videoconferencing and discussion boards. In addition, this section enables users to advertise mentorship programs and to encourage "meetups" for knowledge exchange and collaboration.
(5) *Library/Education Module*: This section will include information portals for educational institutions to advertise, as well as provide direct access to online publications and courses. In addition, this site will include directories of local resources—technology,

entrepreneurship, investment, service providers and government assistance programs—as well as links to RIS initiatives worldwide.

(6) *Marketplace Module*: This module will include a job board, including links to employment agencies. It is a forum for users to post "needed" and "for sale, lend or swap" notices for equipment, lab space, real estate and other items pertinent to RIS development and to knowledge workers.

(7) *News/Announcements/PR Module*: This module will advertise local events and achievements of the local RIS and other RISs. It aggregates technology and geopolitical news that is relevant to RIS development. It is the RISs' face to the outside world. It is important that it is up-to-date and appropriately highlights the local RISs' accomplishments.

(8) *Analytics/Decision Module*: This module includes mapping and data mining tools that show connections among researchers, labs, and patents. It includes individual and group decision-support tools, as well as other analytic tools of management science and operations research.

(9) *Physical Space Support Module*: This module deals with the audio-visual needs of local conventions, museums and visitor information centers. It also encourages meetups in selected physical locations.

(10) *Data Repository/Curation Module*: This module maintains the RIS's history, in words and numbers. Users should have the option of controlling the privacy and sovereignty of the data they provide. A curator checks the authenticity and usable formatting of the information.

Appendix 3 details the functions performed by each module.

11.4 A Tale of Three Cities: Innovative Digital Content in Three RISs

11.4.1 *Portland*

Almost one-third of core high-tech employment in Oregon is in the semiconductor and electronic component manufacturing. Employment in this industry is more than six times more concentrated in Oregon than it is nationally. Other Portland-area clusters include computer systems design, architectural and engineering services, flat-panel displays, sports and outdoor wear (including Nike and Columbia Sportswear), educational technology,[4] and software publishers.

The Technology Association of Oregon (TAO), a grassroots organization, is a very good example of a platform that has facilitated the flow of knowledge and information to support the regional tech industry. For instance, the information and communications technology (ICT) has grown more than 10% per year for the last five years.[5]

The Portland area's technology industry's portal, Techlandia.org, is the most advanced and comprehensive IIRIS that we have found. It bridges the digital divide and creates a safer digital life for the technology community. Techlandia includes interactive content and company profiles. Some of the organizations that Techlandia highlights include Act-on, PDXWIT, and CBTNuggets.

- Portland Women in Tech (PDXWIT) offers community and skills-building events, mentorship, and access to jobs within the Portland area's tech industry. The aim is to maintain the balance in the tech community by empowering individuals in the tech-industry.[6]
- CBT Nuggets offers online educational courses, virtual labs and a training library.[7]
- Act-on offers solutions to engage, convert and measure the marketing channels through automation.[8]

11.4.2 *Austin*

Noticing the startling growth in the number of software companies located in Austin in the 1990s, grassroots leaders formed the initiative organization

---

[4] Northwest Education Cluster. URL: https://nwedcluster.com/.

[5] URL: https://www.techoregon.org/blog/report-shows-nuanced-story-continued-growth-oregon-tech-industry/.

[6] Portland Women in Tech. URL: https://www.pdxwit.org.

[7] CBT Nuggets. URL: https://www.cbtnuggets.com.

[8] Act-on. URL: https://www.act-on.com/.

that evolved into the Austin Technology Council. This organization's networking prowess leveraged the firms' proximity, interaction and access to university expertise to create expanded opportunities for innovators, entrepreneurs and investors. Its visionary influencers transformed Austin into a high-tech hotspot and center for innovation. At the time of the Council's founding, Austin had one technology business incubator and 450 software companies. By 2018, the city boasted 90 incubators and accelerators, 6,500 high-tech companies, 124 venture capital deals[9] and an untold number of other non-profit entrepreneurship support organizations.

With more than 30 years' of experience in attracting technology companies, Austin hosts a true digital supercluster, benefiting from synergies among the electronics of semiconductors, computers, software, music, gaming, energy, transportation and medicine.

High-profile companies such as Dell, Intel, Samsung, Advanced Micro Devices, Bazaarvoice, DXT Technology, Google, National Instruments and Tableau Software call the Austin area home. Given the growth, Information coordination within the region has become a challenge.

The Austin Chamber of Commerce is an independent membership organization that has facilitated the platform for local businesses to launch, expand or create their presence in Austin, and to build a skilled workforce in the region. The website of the Austin Chamber has a list of tabs that offers up-to-date and extensive information on the area's researchers, entrepreneurs, government users and service providers.[10] It is the go-to site for those seeking information about Austin's technology and entrepreneurship scenes. Since 2004, the Austin Chamber has helped bring more than 400,000 jobs to the region.

11.4.3 *Daeduk*

Daeduk, home to 26 government-sponsored research institutes, lies within the Daejeon city limits, but is administered by the Korean national government, not the local city government. Two facilitating agencies, the Innopolis Foundation and the Gyeongbuk Technopolis

[9] Austin Chamber. URL: https://www.austinchamber.com.

[10] Austin Chamber. Opportunity Austin. URL: https://www.austinchamber.com/about/opportunity-austin.

Foundation, have missions that overlap but differ in their official geographic, technological and institutional reach. This, along with very complex government regulations, throws difficulties in the path toward a comprehensive portal for the Daejeon-Daeduk RIS.

The Daejeon Metropolitan City's Korean- and English-language websites are an example of an information system that enables users to learn more about the city. For instance, the website contains portals on the topic of living in Daejeon. This item includes links to pages that contain information on public transportation, the community health center and foreigner-related municipal regulations. In addition to these links, there is a "statistics" section that contains information on the population sizes as of 2018 in Daejeon.[11]

A "Daejeon Metropolitan City" website differs from the one listed above, in that it contains more information related to the research and development industries in Daejeon. For instance, this website includes links to sections that contain information on Daejeon's investment strengths, investment projects, and major industries in the area. In addition to these portals, there is a section that is labeled "Useful Links". This section includes a comprehensive list of the government-invested science and technology research centers and major universities, and information for foreign investors.[12]

The Korea Advanced Institute of Science and Technology is "Korea's MIT". KAIST's bilingual "Startup KAIST" site[13] also appears to be instrumental to the Daejeon-Daeduk RIS.

There is online support for the exchange of personnel and equipment among the national laboratories and research institutes. Connecting research institute technologies and expertise with companies and entrepreneurs is not smooth—in fact, companies and entrepreneurs seeking connection to national laboratory assets must go to a physical office, fill

---

[11] URL: https://www.daejeon.go.kr/dre/DreContentsHtmlView.do;jsessionid=NY2f0NN0YqtUieWpcACeVrDKNs1HJHo0NlT0AdorN7xRBsZny16FkeNcnbSr1CTO.WEB1_servlet_engine1?menuSeq=2734.

[12] URL: http://www.investkorea.org/daejeon_en/info/info01.do.

[13] URL: https://startup.kaist.ac.kr/category/partner-event-en/?lang=en.

out a paper form, and give it to a human clerk, who processes the request with the help of regulation books and technology brokers.

Global Techno Business Center Daeduk emphasizes makerspaces, which we highlighted as part of our "physical space" IIRIS module.

11.5 How do Portland's, Austin's, and Daeduk's IIRISs Compare to the Benchmark?

Austin and Portland do well on the Guide Module. Inquiries will lead to the needed resource. Daeduk's online directories need more customer-orientation. Regarding the Metrics/Evaluation Module, Austin and Portland's resources are updated more regularly on the website, whereas Daeduk's metrics appear in less frequently updated reports. As for a Moderation/Social Engineering Module, all three cities could do a better job on trust-building.

Austin and Portland do a good job on site esthetics, collecting user feedback, and ease-of-use measures, dimensions on which Daeduk could use some improvement.

Collaboration resources are scattered, rather than centralized, in Austin and Portland, but even less well-developed in Daeduk. All three cities offer effective Library/Education Modules.

The Marketplace Modules of all three cities focus solely on job matching. No other buying, selling, or swapping facilities are evident.

The notable shortcoming of all three cities' News/Announcements/PR Module is that they do not appear to reach out to other RIS for knowledge sharing. Their Analytics/Decision Modules offer scarce and scattered analytic capabilities. Yet, all three cities offer good IT and A/V support to their convention centers and other relevant meeting spaces. These services fall under the Physical Space Support Module.

Data Repositories and Curation in Austin and Portland take the form of occasional white papers and academic theses, rather than in an IIRIS module. This is surely partly due to the general future-orientation of RISs, and also due to the reluctance of growing companies' CEOs to acknowledge that their firms could not have flourished without the initial support of public agencies [Phillips, 2006]. On this dimension,

Daeduk does better, with governments habitually publishing annual statistics.

This evaluation does not imply that grassroots RISs do better with information support. We are reporting here only on the three case study cities. The conclusion we draw from this section's discussion is that Austin, Portland and Daeduk, all RISs of a high maturity level, do well with RIS information infrastructure, but could nonetheless do much better.

11.5.1 *Interviews*

Before moving to Thailand, Adam Bates worked in commercialization in the US and with SMEs at the Gyeonggi Small Business Center. He noted Korean organizations' oddly duplicative missions, and their reluctance to share information. He laments that so much of the Korean government research institutes' activity is contract research for large corporations, rather than independently directed research on topics important to Korea and the world society.

Other takeaways from the Bates interview include:

- Company databases are proprietary and those are usually kept by the partners. There is not a lot of overlap between organizations and many times one will find that organizations are very reluctant to share information.
- There are databases that have the basic, vital information for the company as far as what their products are and revenue streams. In the US, there is a treasure trove of information so easily available online; whereas he has never seen anything like that in Korea or Thailand. Having that data is incredibly useful especially when one is doing business.
- After talking with Bates, the differences between how information or knowledge is disseminated in a grassroots-driven RIS (i.e., his past work in Austin) versus a government-driven RIS (i.e., his work in Korea and Thailand) became clearer. These key differences include a higher level of data or idea protection and less information-sharing across economic, social, political and institutional entities in a government-driven RIS than that of a grassroot-driven RIS.

An interview with Tom Stephenson and Webb Johnson revealed that:

- Typically, the main force motivating the development of websites within innovation systems will come from a single individual who takes on the challenge of promoting connectivity across different areas. In other words, it is personality driven and requires a single individual to help move knowledge within a community.
- When talking about community-based initiatives, things often become more challenging because it is about the preservation of content. For instance, if one goes to a website and the links are broken and/or not updated, then there is an immediate discounting of the network effect.
- While economic growth is centered around startups, it is interesting because there are differing views on what will get people to contribute to economic growth.
- One of the ways to recruit scientists to come to areas that are not advanced RISs is the cost of living, the affordability of extracurricular activities for children and the lack of traffic.
- If there is any "secret sauce" to any of these things, it is the committed individual; whether they are supported by the government, private, etc.; that is how it works.
- There is a fear that comprehensiveness makes it impossible to deliver.
- Overall, if one is looking for a national laboratory job, one is not terribly entrepreneurial.

The Stephenson-Johnson interview highlighted the fact that a bottom-up IIRIS, depending on the initiative of one or more committed individuals, may break down as those individuals move on to other endeavors. In contrast, top-down government-driven IIRISs tend to persist.

The latter interview reinforced the idea that entrepreneurs have a tolerance for risk while government workers generally do not. That means the presence of government laboratories in the region is not, generally, a spur to local entrepreneurship. However, in some regions, young knowledge workers do not expect to work in government laboratories for their entire career. Thus, the IIRIS may serve entrepreneurship by informing lab workers of market needs and opportunities.

Another of their points was that 'perfect' is the enemy of 'better'. A region hoping to affect all the functions and modules in our model may never do so. For this reason, we have included in this chapter a template map (Table 11.2) of IIRIS functions against RIS maturity levels.

Also emerging in this interview was the importance of the IIRIS highlighting the region's quality of life and cost of living.

The interview with Kevin Koym, founder and CEO of Austin's Tech Ranch, highlighted the role of local culture in determining propensity to share information across organizations. Koym emphasized the importance of pictorial and video content on RIS portals (due to the fact that many entrepreneurs are dyslexic, and thus becoming entrepreneurs because they cannot deal with the corporate setting's text-heavy communications), supplemented by face-to-face meetings of entrepreneurs. Also valuable was Koym's view on the velocity of information (frequency of portal updating) as it feeds entrepreneurs' needs for "learning on demand".

Han Sic Cho of the city of Daejeon noted the two dozen government research institutes located in Daeduk were placed there for ease of logistics, without a plan for cooperative synergy among the institutes. Though the institutes share equipment and personnel, the arrangements for connecting government lab expertise and intellectual property with companies and entrepreneurs remain awkward. The distinct but overlapping jurisdictions of Daejeon, Daeduk, the Innopolis Foundation, and the Gyeongbuk Technopark, and the various regulations that govern them, seem to make a central, "go-to" portal for the metropolitan area a quite difficult proposition.

Raimund Bröchler of INTRASOFT International noted that the next step after assuring data security is assuring "data sovereignty", i.e., that data owners control the access and use of their own data.

11.6 Information Support for RIS and for Smart City: Where Do They Overlap?

This section is based on our conversations with Hung Song, retired CEO of the KT-Cisco joint venture that provided the hardware, software, and control under-layer for the Songdo (Incheon, South Korea) smart city, and with Sheridan Tatsuno, who authored the first book on Technopolis [Tatsuno, 1986] and just published an article[14] on smart city in *Forbes* magazine [Tatsuno, 2019].

Smart city uses Internet of Things (IoT) to enhance the quality, efficiency and revenue-generating capacity of urban services. These services serve residents and businesses. Most of the services, such as smart streetlamps, smart parking meters, and ubiquitous sensors are not of direct relevance to attracting technology companies or technology entrepreneurs.

Smart city infrastructure may reduce municipal costs (and hence taxes) in the long run, though according to Song, they have never yet proven to do so. In the short run, a city's upfront expenses in acquiring the infrastructure requires raising taxes. This makes a smart city less attractive to cash-strapped startup ventures.

Smart city plans originated with the IT provider firms; the firms sell their visions to the municipalities, and generally implement them with minimal or no input from residents. This contrasts with the more organic growth of a technopolis or RIS, in which creative people design and experiment with their own ideas for communication, transport, learning and so on. Tatsuno agrees that smart cities increase corporations' control over ordinary citizens' daily lives.

In short, the smart city concept is somewhat antithetical to the RIS concept.

However, there are two areas where the smart city and the RIS share an interest in information infrastructure. One is IoT. RIS firms use IoT for manufacturing, logistics, and in products. A smart city located in an RIS enjoys much local expertise in IoT. On the other hand, educated people working in the RIS are wary of the "Surveillance State" and use IoT devices sparingly. Song, for example, lives in a smart apartment but never uses its smart features.

The second area of overlap has to do with quality of life (QOL). QOL is essential for attracting knowledge workers, who are generally mobile and can live and work anywhere. To the extent that smart city equipment enhances local cultural, educational, and recreational opportunities, and responsive government, the interests of RIS and smart city coincide.

---

[14]Link to the Forbes Article, https://www.forbes.com/sites/cognitiveworld/2019/03/03/the-evolution-of-smart-cities/?sh=3aeac7365df5.

11.6.1 *Analysis and interpretation: Why do RISs not have comprehensive information support*

Government-driven RISs and grassroots-driven RISs face similar obstacles to comprehensive information support, but experience these obstacles in different ways. Table 11.1 sums up this comparative analysis.

In nascent RISs, a single initiative organization can drive progress; for example, the IC^2 Institute in Austin served this function for some years. As the RIS matures, additional facilitating organizations spring up to address specialized needs within the RIS. Examples might include female entrepreneurship, the gaming industry or social entrepreneurship. Each new organization keeps its data close to the vest, in order to differentiate itself and attract funding. This is the "maturation/fragmentation" item seen in Table 11.1.

A centralized IIRIS that fails to show new and vital information, or that appears not to have been updated for ages, will lose viewers.

Under Table 11.1's "strategic" considerations, government organizations with different and strictly defined missions will find little common ground for building a joint information system. A metro area should have an economic development strategy. If the strategy is too fuzzy, its information system will show poor focus. Industry 4.0 capability depends on IoT, which in turn depends on 5G cellular/wireless network. Current news stories show that weather forecasters, astronomers, and health officials are wary about the impact of 5G.

An RIS includes many actors. Understanding the needs and agendas of each is a task in which not everyone is skilled. Yet, the IIRIS must meet these diverse needs—and have an attractive interface—in order to retain users. The interview with Koym highlighted the importance to entrepreneurs of non-text information formats.

Reluctance to share data to a common platform may stem from competitive pressures or from government regulation. It may also stem from poor social capital in the region, a lagging transformation to digital operation (as shown by the persistence of paper forms in our Korean example), or from various actors differing views of what an RIS is and how it should benefit the region.

Table 11.1. Summary of obstacles to IIRIS.

	Source	Grassroots RIS	Government-led RIS
Dynamics			
Maturation/fragmentation	3, 9	√	
No continuity	2	√	
Infrequent updates	2, 3	√	√
Strategic			
Differing missions and remits	5		√
No coherent RIS strategy	5, 9	√	√
Concerns about 5G	7	√	
Customer (dis)orientation			
Poor understanding of RIS actors' needs	5	√	√
Poor user interfaces	9		√
Little video content	3	√	√
Secretive			
Proprietary data	1	√	
Privacy concerns	4	√	
Reluctance to give up data sovereignty	6	√	
Regulations prevent information sharing	5		√
Cultural			
Poor collaboration culture	1, 3	√	√
RIS is slow to adopt digitalization	5	√	√
Differing underlying philosophies of RIS	2	√	√
Other content issues			
No committed moderator	2, 9	√	√
Insufficiently comprehensive content	2	√	√
Lack of compelling content	9	√	√
No killer app	8	√	√
Unable to link to government census data	1		√

Sources: (1) Bates interview; (2) Stephenson-Johnson interview; (3) Koym interview; (4) Song interview; (5) Cho interview; (6) Bröchler interview; (7) Davis [2020]; (8) Townsend *et al.* [2009]; (9) Authors' own observations.

"Other content issues" include the very important considerations of a committed moderator, and a killer app that smaller organizations in the region, perhaps, cannot afford individually.

11.7 Conclusion

Like all 21st century organizations, regional innovation systems (RISs) require information support. This chapter has outlined the best-practice functions to be provided by RIS information support, grouping these functions into several modules, and calling the result an Information Infrastructure for Regional Innovation Systems, or IIRIS.

Except for Austin, Texas, and the Oregon/southern Washington region, which come close, we found no RISs that enjoy the full, ideal range of IIRIS functions. Even in Austin, with its highly developed Chamber of Commerce site, information resources are fragmented among many NGOs. Surprisingly, even the report [Ottinger and Trapido, 2019] on the vital Seattle innovation ecosystem mentions no comprehensive, or even partial IT support for the RIS.[15]

Answering the question posed in this chapter's title, we found 21 reasons (Table 11.1) why centralized and comprehensive information support is not found in the typical RIS. Overcoming these obstacles would require cultural and regulatory reforms.

11.7.1 *Limitations of the Research*

Though we found our interviewees forthcoming and frank, interviewees tend to put their best foot forward, and this bias is always present in qualitative research.

Identifying grassroots organizations with the Occident, and government-driven RISs with the Orient, is valid only in a general way. This project's focus on the western grassroots RISs of Austin and Portland, and the top-down Korean RIS, was happenstance.

---

[15] A possible explanation for this is that so many of Seattle-area innovators are WeWork tenants, availing themselves of WeWork's internal portal.

Table 11.2. Template for prioritizing IIRIS module development.

	Maturity level 1	Maturity level 2	Maturity level 3	Maturity level 4	Maturity level 5
"Guide" Module					
Metrics/Evaluation Module					
Moderation/Social Engineering Module					
Collaboration Module					
Library/Education Module					
Marketplace Module					
News/Announcements/ PR Module					
Analytics/Decision Module					
Physical Space Support Module					
Data Repository/Curation Module					

11.7.2 *Recommendations*

Successful information support for RISs requires cultural and regulatory measures. Austin's story shows that a culture of information-sharing, so helpful for RIS development, may not persist over the long run. Daeduk's story illustrates that regulatory reform, and perhaps a consolidation of institutes and agencies, would facilitate greater information sharing, internally and with the general public. In both locations, a protocol similar to the Creative Commons[16] may make constituents feel more comfortable with data sharing.

Because each hopeful techno-region has its own needs, resources, and budgets, we present Table 11.2 as a tool for prioritizing IIRIS

[16] URL: https://creativecommons.org/.

development, but leave the cells blank. RIS initiatives, developers and moderators may use Table 11.2 for planning purposes.

References

Belanche, D., Casaló, L.V, and Flavián, C. (2012). Integrating trust and personal values into the Technology Acceptance Model: The case of e-government services adoption, *Cuadernos de Economía y Dirección de La Empresa, 15*(4), 192–204. URL: https://doi.org/https://doi.org/10.1016/j.cede.2012.04.004.

Constantinides, P., Henfridsson, O., and Parker, G.G. (2018). Introduction—Platforms and Infrastructures in the Digital Age, *Information Systems Research*, *29*(2), 381–400. URL: https://doi.org/10.1287/isre.2018.0794.

Davis, D. (2020). How Pollution Aggravates the Impact of Coronavirus, *US News & World Report,* March 16. URL: https://www.usnews.com/news/best-countries/articles/2020-03-16/commentary-pollution-in-italy-china-and-iran-worsens-the-coronavirus-impact.

Dieter, E. (2014). Trade and innovation in global networks: Regional policy implications, in: van der Giessen, A., Stolwijk, C., and Leijten, J. (eds.), *Can policy follow the dynamics of global innovation platforms?* Hertogenbosch/Den Bosch, The Netherlands, 6CP.

Egbetokun, A., and Niosi, J. (2013). Building national and regional innovation systems: Institutions for economic development, *Science and Public Policy*, *40*(2), 279–280.

Lee, C., Bae, Z.T., and Lee, J. (1994). Strategies for Linking Vertical Cooperative R&D to Commercialization in Korea, *Journal of Product Innovation Management*, *11*(4), 325–335. URL: https://doi.org/10.1111/1540-5885.1140325.

Lee, Y., and Kim, H. (2016). Daedeok innopolis as global science scape: Its possibilities and limitations, *Global Science Spaces working paper No. 3*. URL: www.globalsciencespaces.org.

Leydesdorff, L., Porto, I., Comins, J.A., and Phillips, F. (2019). Synergy in the knowledge base of US innovation systems at national, state, and regional levels: The contributions of high-tech manufacturing and knowledge-intensive services. *Journal of the Association for Information Science and Technology (JASIST), 70*(10): 1108–1123. DOI: 10.1002/asi.24182.

Oh, D.S., and Phillips, F. (eds.). (2014). *Technopolis: Best Practices for Science & Technology Cities.* Springer.

Ottinger, R.J., and Trapido, D. (2019). *Greater Seattle Innovation Ecosystem Report*, January, pp. 1–70.

Porter, M.E. (1998). Clusters and the new economics of competition, *Harvard Business Review*, 77–90.

Phillips, F. (2006). *Social Culture and High Tech Economic Development: The Technopolis Columns*. London: Palgrave.

Phillips, F. (2008). The Godfathers: Characteristics and roles of central individuals in the transformation of Techno-Regions, *Journal of Centrum Cathedra*, *1*(2), 12–27.

Phillips, F., Alarakhia, S., and Limprayoon, J. (2014). The Triple Helix: International Cases and Critical Summary, in: Oh D.S., and Phillips, F. (eds.), *Technopolis: Best Practices for Science & Technology Cities*. Springer.

Phillips, F., Matheson, B., and Thapa, U. (2019). *Technology Commercialization in Regional Innovation Systems, with a Focus on Information Infrastructure*. Korea Institute for Science & Technology Information (KISTI).

Schot, J., & Steinmueller, W.E. (2018). Three frames for innovation policy: R&D, systems of innovation and transformative change, *Research Policy*, *47*(9), 1,554–1,567. URL: https://doi.org/https://doi.org/10.1016/j.respol.2018.08.011.

Srinivasan, A., and Venkatraman, N. (2018). Entrepreneurship in digital platforms: A network-centric view, *Strategic Entrepreneurship Journal*, *12*(1), 54–71. URL: https://doi.org/10.1002/sej.1272.

Tassey G. (2019). Regional technology-based economic development: Policies and impacts in the U.S. and other economies, *Annals of Science and Technology Policy*, *3*(1), 1–141. URL: http://dx.doi.org/10.1561/110.00000012.

Tatsuno, S. (1986). *The Technopolis Strategy: Japan, High Technology, and the Control of the Twenty-First Century*. Prentice Hall.

Tatsuno, S. (2019). The evolution of smart cities, *Forbes*, March 3. URL: https://www.forbes.com/sites/cognitiveworld/2019/03/03/the-evolution-of-smartcities/#353187405df5.

Townsend, A., Pang, A.S.K., and Weddle, R. (2009). *Future Knowledge Ecosystems: The next twenty years of technology-led economic development*. Institute for the Future (IFTF Report No. SR-1236).

Appendices

A.1 List of Interviewees

- Adam Bates, Mahasakharam University, Thailand. Formerly Program Manager at the IC^2 Institute of the University of Texas at Austin.
- Raimund Bröchler, Senior Research and Innovation Manager, INTRASOFT International S.A., Luxembourg, and Vice President of International Network of SMEs (INSME).
- Han Sic Cho, Director International Relations, Daejeon Metropolitan City.
- Webb Johnson, RIS activist, Albuquerque.
- Kevin Koym, CEO, TechRanch, Austin.
- Gina Phillips, Seattle, USA. Vice President of WeWork.
- Junseok Seo, Senior Advisor for Policy at the Innopolis Foundation.
- Hung Song, Songdo, Korea, former president of Cisco-Korea Telecomm JV managing the Songdo smart city project.
- Tom Stephenson, investor, Albuquerque.
- Sheridan Tatsuno, Santa Cruz, California, author of *The Technopolis Strategy*.

A.2 Interview Questionnaire for RIS Leaders

Thank you for speaking with me. There has been a lot of research about regional innovation systems, but very little about the information infrastructures that support RIS development. I am referring to information resources that all actors in the RIS—investors, entrepreneurs, nonprofits, universities, large companies, journalists, etc.—can access most quickly and conveniently to smooth the path of local innovation.

(1) As you are a leader in your region's innovation system, I want to ask you about the information resources that support its growth. I

have sent two documents to you in advance of this interview. Have you had a chance to look at them?

(2) Is there a primary "go-to" website that all innovation actors in the <insert region name, e.g., Portland and Oregon>, and all newcomers, know about and use?

(3) What organizations other than yours maintain websites and other innovation information resources for the region? Please name the main ones so that we can examine their sites.

(4) What is the state of their connectivity? Do these organizations link their sites to each other, or are they siloed? What are the forces leading to linking or to silo-ing?

(5) Currently, which local organization maintains the information system most useful to innovators? What is the most used information system available?

(6) What is these systems' range of collaboration mechanisms? For instance, are there chat rooms, videoconferencing, discussion boards, job boards, etc., that enable people to find and interact with one another online? To what extent do these mechanisms match the "outline" that I sent to you? Do you have suggestions for information functions that should be added to the outline, whether your region currently offers them or not?

(7) What are the motivating forces for the development of different website(s) for a regional innovation system (RIS)?

(8) What motivates people to use the systems?

(9) How do you imagine newer ITs improving the information support for the RIS? For example, IoT, Blockchain, AI?

(10) Using the Technopolis Maturity chart that I sent to you, how would you assess your region's innovation maturity level, on the of 1 to 5 scale?

Thank you so much for your help and direction this morning.

A.3 Detail of an Ideal Information Infrastructure for a Regional Innovation System

"Guide" Module (Module 1)	• Suggests IIRIS navigation paths for: – Researchers – Entrepreneurs – Larger companies – Government users – Research/industrial parks – SMEs – Service providers • Lawyers • Accountants • Commercial realtors • IT services • Etc. – And for RISs of type • Market • Social democratic • Continental western Europe • Meso-corporatist – And for RISs of maturity levels 1, 2, 3, 4, 5
Metrics/Evaluation Module (Module 2)	• Maturity level assessment • Tracking progress against role models • Tracking progress against goals
Moderation/Social Engineering Module (Module 3)	• Site esthetics • Ease-of-use measures • Active moderation • Trust-building • Ensure messages are not on constituents' spam lists • Outward messaging • Suggesting connections • Ensuring completeness & timeliness • Celebrating successes • Collecting user feedback • Sharing tools with other RISs • Cross-linking with other relevant local sites • Generating reasons to participate – For all user types

Collaboration Module (Module 4)	• Chat rooms • Google Docs-type collaborative authoring • Blockchain tracking of research contributions • Videoconferencing • Discussion boards • Bulletin board – Seeking expertise, collaborators, references • Mentorship matching – Big corp., VC, SME, Startup • Arranging "meetups"
Library/Education Module (Module 5)	• Directories – Companies – Organization membership – Individuals prominent/active in RIS – Pertinent regulatory agencies and regulations – Investor directory • Who has funded what • Curated YouTube video links • Virtual incubation, online pitch contests • Subscriptions to market/environmental scanning services • Internship programs • STEM, entrepreneurship, liberal arts curriculum materials for K-12 teachers, students. Links to MOOCs • Technology forecasts • Meeting minutes
Marketplace Module (Module 6)	• Job board • Equipment swap • Seeking – Partners – Customers – Suppliers
News/ Announcements/ PR Module (Module 7)	• Event planning tools • Event announcements & registration • Digests of RIS news worldwide • Visitor tracking / CRM • Newsletters for external distribution

(*Continued*)

Analytics/Decision Module (Module 8)	• Identifying research centers of excellence • Goal-setting – Finding role model regions • Group decision rating/ranking tools • Patent search/analysis • Mapping tools – Patent forward/backward citations – Co-authorships – Other researcher networks
Physical Space Support Module (Module 9)	• Physical site-based electronic support for: – Conventions – Public relations – Museums – Dioramas – Makerspaces – Other equipment shared across companies – Meeting rooms
Data Repository/ Curation Module (Module 10)	• Preserve data • Update data • Clean data • Check data accuracy

Part 3

Best Practices

© 2023 World Scientific Publishing Company
https://doi.org/10.1142/9789811257186_0012

CHAPTER 12

Smart City Strategy for Building an Innovation Platform: The Daedeok Innopolis

Sung-Kwang Yang

Innopolis Foundation, Daejeon, Korea
Email: skyang@innopolis.or.kr

12.1 Introduction

The Daedeok Research Complex which had been planned since 1973 was completed in 1992 by relocating government-funded research institutes and educational institutes like the Korea Advanced Institute of Science and Technology (KAIST) scattered around the capital area, and by bringing in central research centers of conglomerates. Afterwards, the Daedeok Research Complex has dutifully played a vital role in the nation's economic growth by supplying core technology and high-quality human resources in the 1980s and 1990s.

The Daedeok Research Complex was converted into an innovation cluster with an added industrial zone at the rear of the existing research and development (R&D) Zone in 2005, in accordance to changed national needs in the science and technology environment, which has been expanded to five regions nationwide until 2015. As of the end of 2018, 2,046 institutes including government-funded research institutes, universities and companies have moved into this special zone with 7.5 million workers, and this zone has developed into a pivotal center for science and technology in Korea, while owning 208,000 domestic patents and 208,000 overseas ones applied.[1]

The R&D Special Zone is now making an attempt of innovation such as the spread of open innovative culture and the empowerment of commercialized technology in order to play a proper role as a hub for national innovative growth in the 4th Industrial Revolution era. For this purpose, a structural innovation is being carried forward by strengthening a connection between the Daedeok R&D Special Zone (Daedeok Innopolis) with its accumulated maximum research resources and the rest of special zones, and by newly designating 'small but strong R&D Special Zone' while focusing on one or two key institutes.

The Daedeok Innopolis is also striving for a change to cope with the technological innovation, which utilizes 4th Industrial Revolution technologies like AI, Big Data and robots. A "renovation Master Plan" is being established to create a responsive innovative ecosystem where startups and commercialized technology are invigorated by adding activity space for companies to the integrated zone of research centers and by combining a concept of smart city to build an innovative environment.

This study aims to look into the characteristics of the 4th Industrial Revolution era, its influence on human life and society, and the relevant changes in the global market environment, and to briefly introduce efforts to change by innovators in Daedeok Innopolis and the policy direction of building an innovative ecosystem of smart city in the future.

12.2 Arrival of the 4th Industrial Revolution Era and Change of Market Environment

12.2.1 *Characteristics of the 4th Industrial Revolution era*

The most important characteristics of the 4th Industrial Revolution era are broadly classified into Super-intelligence, Hyper-connectivity and Virtualization (see Figure 12.1). Artificial Intelligence (AI) and Internet of Things (IoT) are key technologies that lead to these characteristics. In

---

[1] Korea Innovation Foundation (www.innopolis.or.kr) statistical data of the special zone.

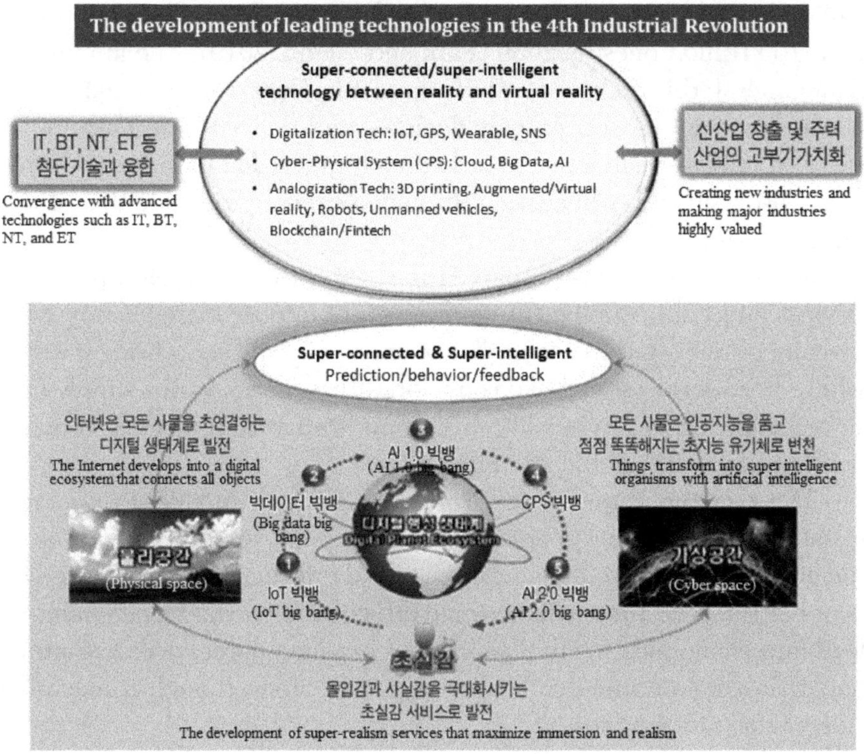

Figure 12.1. Key leading technology concepts of the fourth industrial revolution.
Source: Kim, 2017, p. 34.

the IoT environment, objects equipped with sensors exchange real-time data through the Internet.

Though the devices connected to the Internet exchanged information in the past, human intervention was needed at that time. However, in the IoT era, the objects connected to the Internet can exchange information without the help of humans. Countless objects like refrigerators, boilers and cars are also networked to communicate with each other.

The data produced by the sensors contain such abundant and precise information that human senses cannot detect. More than one trillion sensors are expected to be used worldwide by 2020 and more than 100 trillion ones by 2030 [Park and Glenn, 2019]. The amount of data generated is enormous. With these Big Data, Internet-connected AI analyzes, processes, judges, decides, and delivers necessary measures. It judges much more smartly than humans do by using more precise sensors and AI, while going through a similar process to that of the human brain.

A house turns into a Smart Home with interconnected personal devices and home appliances, and a factory into a Smart Factory with machines connected to each other. A smart city will soon bring a world where hardware systems like transportation, energy, water supply and sewage, and social services like education, welfare and medical care are operated in an integrated and efficient way.

The fact that super-speed Internet network and Wi-Fi are well installed and that many people use them does not automatically result in the creation of wealth. Even when data and information are scattered all over the Internet, there will be some difficulty in creating a new value to lead into economic growth if we cannot produce the services and products that are combined with idea and technology, though it may give convenience to our lives.

12.2.2 *Market change by innovative technology*

As previously mentioned, the world is rapidly changing due to the emergence of innovative technology such as AI, Big Data and robots. In accordance with it, our lives, lifestyles, and markets also swiftly change (see Figure 12.2). The integration of machines, product facilities, warehouses and logistics system is expected to bring about a world that will be precisely controlled by AI in the near future. Big Data, also allows AI to lead market change by accurately predicting people's behavior patterns and demands, and by eliminating product inventory with real time match between distribution and supply.

Customer needs differ depending on their acceptance level of change and accordingly, the market changes. The life patterns of customers are changing, as are their demands for quality products and services. Companies survive only if they accurately understand the

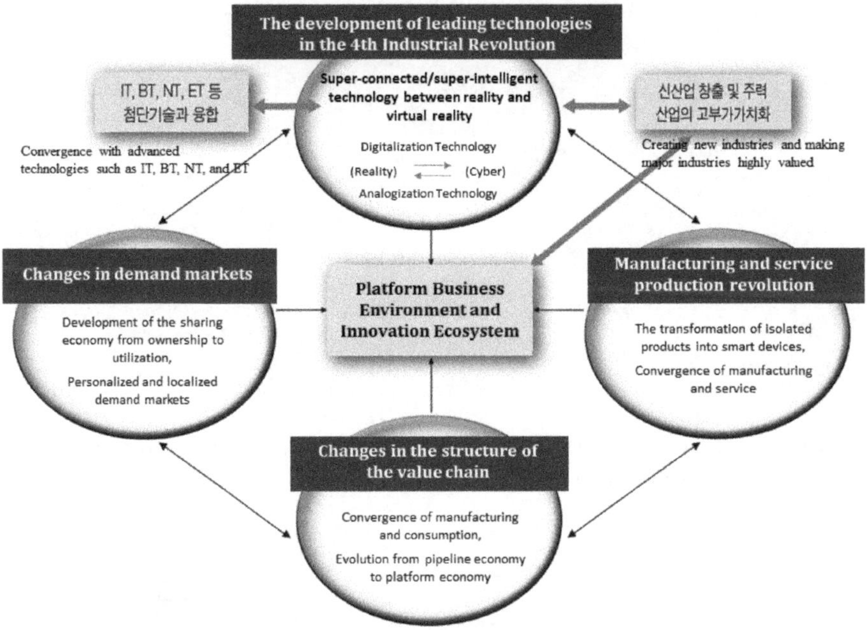

Figure 12.2. Features and effects of the fourth industrial revolution.
Source: Kim, 2017, p. 36.

direction and speed of these changes and prepare in advance. The services and products which companies produce gain the competitiveness only when they meet consumers' standards in a smart way.

The global economy is also rapidly reorganizing into a network-based platform business and sharing economy (see Figure 12.3). In the traditional industry, the value is created by stages in virtue of value chain in which company designs, produces and sells products, and consumer purchases them. However, this traditional method of production that analyzes the past consumption trends and determines the amount of production inevitably results in inventory. However, platform business does not bring about inventory as producers and consumers create value by interacting with each other. Uber and Airbnb create value without owning cars and real estates. Alibaba, the world's largest company, does not have goods in stock through connected platform between producers and consumers. Therefore, platform-based companies, though latecomers, have resulted in overwhelming traditional ones in

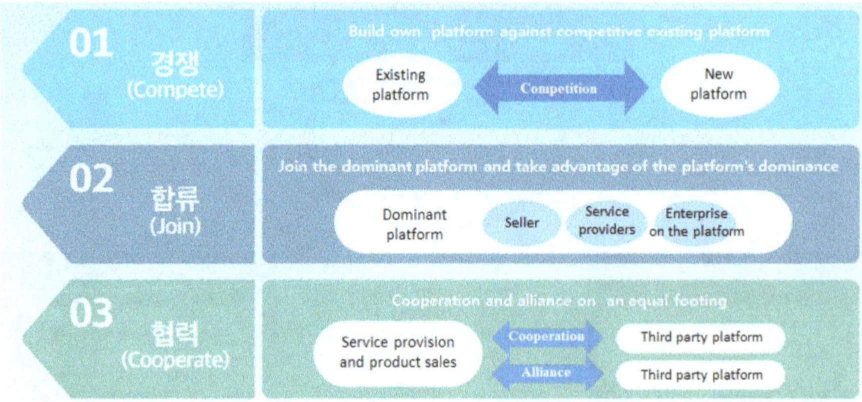

Figure 12.3. Competitive and collaborative strategies for platform business.
Source: Samjong KPMG Economy Research Institute, 2019.

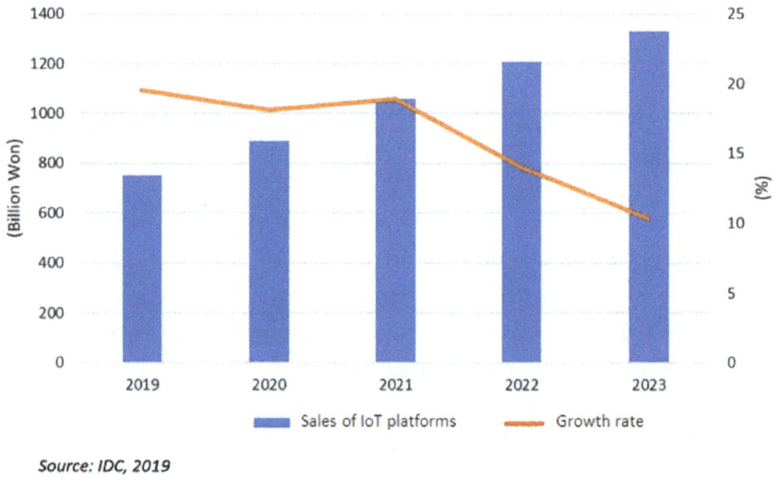

Figure 12.4. Forecast for South Korea's IoT platform market, 2019–2030.
Source: IDC, https://www.idc.com/getdoc.jsp?containerId=AP47358621&pageType=PRINTFRIENDLY.

the same market. The Korea international data corporation (IDC) says that the volume of domestic IoT platform market reached 754 billion won in 2019, which saw an increase of 19.5% from the previous year (see Figure 12.4).

The corresponding market is expected to reach 1.338 trillion won at the annual growth rate of 16.1% until 2023.[2]

12.2.3 *Convergence acceleration between ICT-based industries*

Another recent change in the market is that the convergence between ICT-based industries is accelerating. The convergence between industries is rapidly progressing with the disappearance of the characteristics of individual industries and the collapse of their boundaries, as the individual industries which have been maintained and managed almost independently are converging through ICT. The convergence between different industries has begun to emerge within the convergence of the telecommunication industry based on the former's technology and network. The scope of convergence is extending beyond just a telecommunication field into other industries.

In this environment, the value chains in industry field are connected to one another to form a new market while getting out of the way of how single products compete. For example, in the smart mobility sector, companies that deal with sensors, self-driving cars, electric vehicles, sensors related to the vehicle-sharing business, battery, communication, maps, security, Big Data and AI compete with one another or make alliances to cooperate and work together.

Based on ICT, the active inter-industry convergence will result in changes of a horizontal convergence focused on a platform beyond that of unit services and of the swift collapse of barriers and boundaries between different industries in the near future.

12.2.4 *Spread of smart city strategy utilizing ICT technology*

The technologies of Super-Intelligence, Hyper-Connectivity, and Virtualization in the 4th Industrial Revolution era have a great effect on not

[2] IDC, 2019, https://www.idc.com/getdoc.jsp?containerId=prAP46180120.

only objects and industries but also on the space and environmental sectors. Smart cities are spreading fast in order to preemptively respond to the 4th Industrial Revolution and to create new growth engine, along with the efficient solution to the urban problems such as the lack of resources and infrastructure, traffic congestion and energy shortage resulting from urbanization [The 4th Industrial Revolution committee, 2018].

The concept of the smart city is defined and utilized in many different ways depending on levels of the economy and development of each country, and on the urban situations and conditions. Thus, there are different approaching strategies by country. In general, it can be defined as 'a town model that can solve a variety of urban problems and improve quality of life by incorporating new technologies such as ICT and big data into the city' [Ministry of Land, 2019, p. 6] (see Figure 12.5). Particularly, rising countries like Asia are carrying forward smart city policies led by the public for strengthening country's competitiveness to solve the problem of rapid urbanization and to plan a revitalization of the economy [The 4th Industrial Revolution committee, 2018].

The market volume of a smart city is expected to grow at an annual average rate of 18.4% to 617.2 billion dollars (approximately 692 trillion won) in 2023 [Markets and Markets, 2019] (see Table 12.1).

Korea has implemented a Ubiquitous City (U-City) formation plan since 2008 to provide urban services regardless of time or place, by utilizing technology of information and communication as well. In 2017, the Smart City Act was newly enacted and the 3rd Comprehensive Plan for Smart City was established.[3]

Recently, smart cities which not only solve urban problems at low cost through the efficient utilization of the existing infrastructure by making use of ICT technology, but also have various strategies and

[3] Ministry of Land, Infrastructure and Transport, 2019 Third Smart City Comprehensive Plan 2019–2023, p. 16.

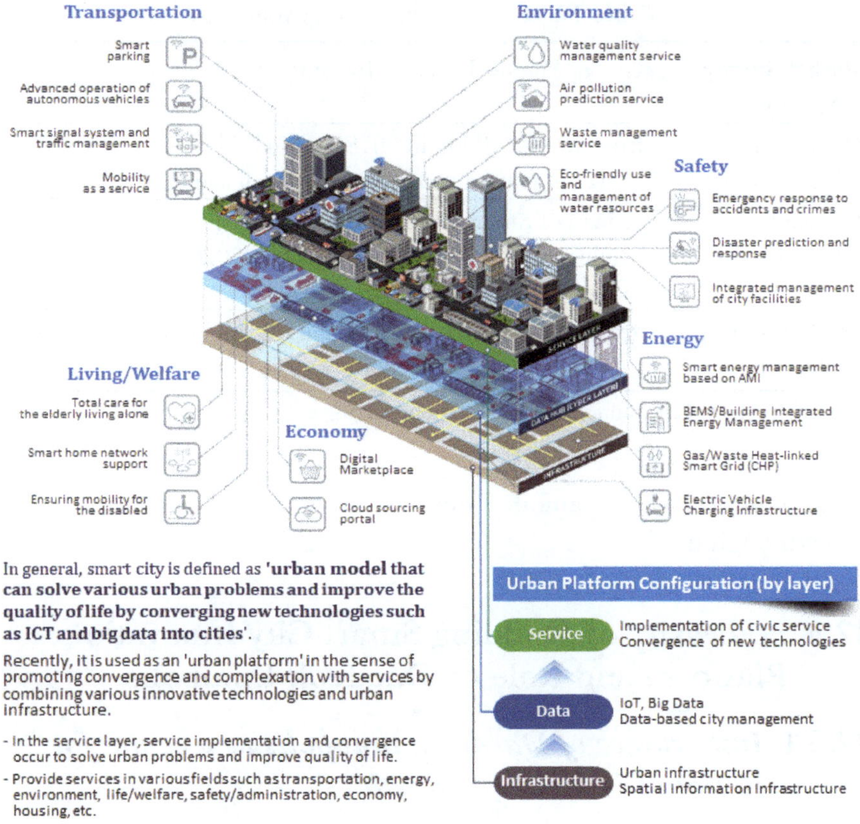

Figure 12.5. The concept of a smart city.
Source: Ministry of Land, Infrastructure and Transport, 2019, p. 6.

contents such as (1) urban platform (data hub), (2) Living Lab and (3) construction of demonstration city are emerging.

Particularly, a strategy to form innovative ecosystems for smart cities has appeared, which aims to find innovative growth engines in countries and regions in connection with smart cities. Research and development (R&D) for demonstration is to be carried out by using the smart city as a test bed for new core technologies that can be

Table 12.1. Smart city market volume.

Market Survey Agency	Global Market Forecast for Smart City
Markets and Markets ('19.1)	from $3,080 billion in 2018 to $617.2 billion in 2023 overall market growth at an annual average of 18.4%
Navigant Research ('17.4)	growth of Smart city service market will be from $93,5 billion in 2017 to $225.2 billion in 2026 (about 113 trillion won)
Frost & Sullivan ('17.10)	growth of global Smart city market will be $2.1 trillion by 2025 (about 2,400 trillion won)

Source: Ministry of Land, Infrastructure and Transport, 2019, p. 16.

implemented in cities among new future technologies for innovative growth engines.

12.3 Necessity of Building Smart City Innovation Platform and Role of Daedeok Innopolis

12.3.1 *Importance of Daedeok Innopolis as a center for national innovative growth*

An effort to construct an innovative cluster based on regional innovation competency first carried forward the Daedeok Science Park located in Daejeon in 2005. The Daedeok Science Park which was completed in 1978 has been transformed from a Science Park focused on research institutes into an R&D special zone by adding a rear district equipped with test facilities and manufacturing operations for companies. The national R&D special zones later extended to five regions including Gwangju, Daegu, Busan and north Jeonla province, etc., and they have now grown into regional innovative hubs with 1.95 million workers at 5,765 institutes including government-funded research institutes, universities and companies.[4]

[4] Korea Innopolis Foundation, 2019. Overview of Innopolis across the country. URL: https://www.innopolis.or.kr/board?menuId=MENU00662&siteId=null.

Overview of Daedeok Innopolis	
• (Designated) July 2005 • (Spatial range) Daejeon Yuseong-gu, Daedeok-gu • (Area) 67.8 km^2 • (Specialized field) IT convergence, Nano fusion, Bio-medical, Precision equipment, etc. • (Main agency) KAIST, CNU, National research institutes (ETRI, KAERI, KRISS, KIMM, KRICT, etc.), Large Enterprise Research Institutes (LG chemical institute, LG energy solution research institute, SK innovation research institute, etc.)	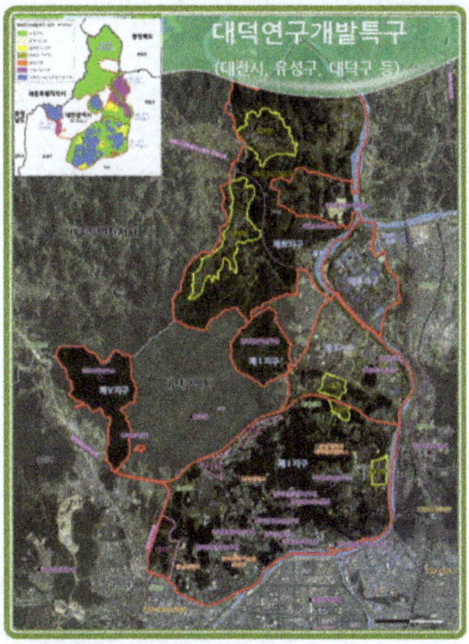

Figure 12.6. Overview of Daedeok Innopolis.
Source: Ministry of Science and ICT, 2018, p. 77.

Among them, the Daedeok Innopolis (see Figure 12.6) is located in Yuseong-gu, Daejeon Metropolitan City, with a total area of 67.8 square kilometers. As of 2018, 2,046 institutes in total are occupied, including 26 funded research centers, seven educational institutes, 12 national and public research institutes and other research centers, 27 government and public institutes, 1,948 companies, and 26 other non-profit institutes.

As of the end of 2018, there are 75,145 workers in Daedeok Innopolis of whom 35,521 (47.3%) are working in research and technology department, and 39,621 (52.7%) are in production and administration.

Daedeok Innopolis is currently continuing to grow. The number of businesses increased from 687 as of 2005 to 1,948 in 2018, at the annual average rate of 14.1%, and the number of employees grew from 6,800 to 400,000 (see Figure 12.7). The figure in the R&D department also increased from 1.8 trillion won in 2005 to 8.3 trillion won in 2018 (see Figure 12.8). These numbers have resulted in about 2.08 million domestic patents and about 9.6 million overseas patents until now (see Figure 12.9).

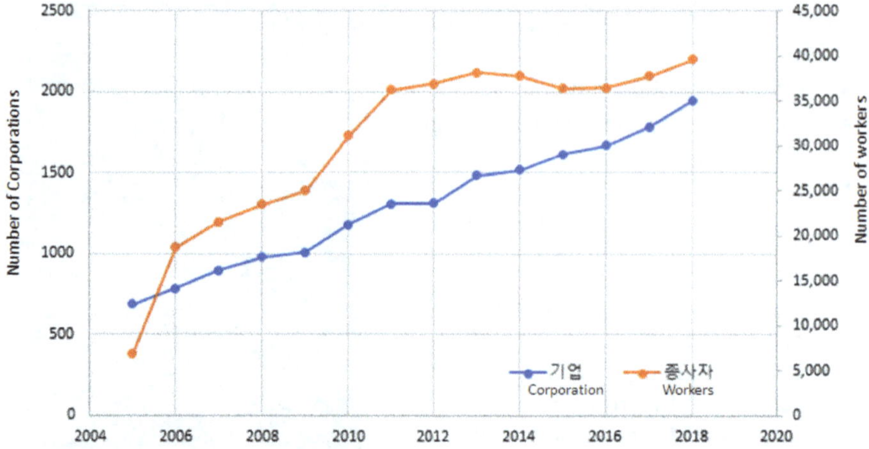

Figure 12.7. Current status of enterprises and employees in Daedeok Innopolis.

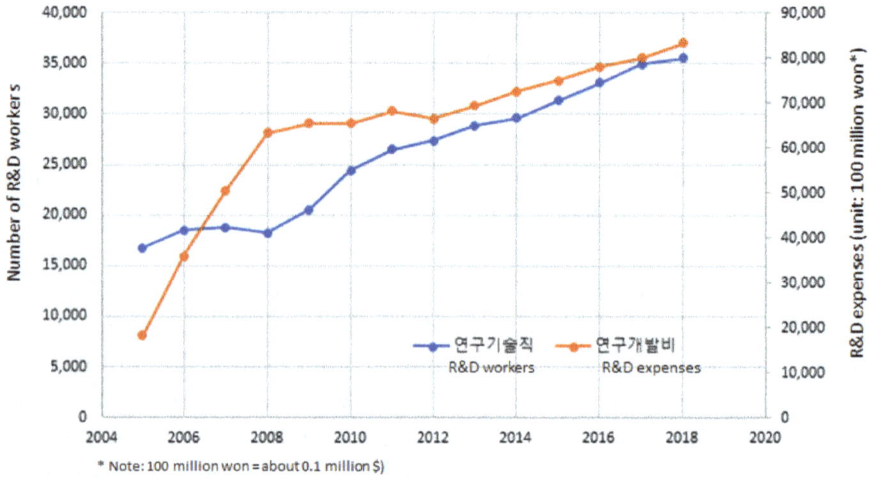

Figure 12.8. Current status of R&D expenses and employees of research institutes in Daedeok Innopolis.

12.3.2 *The necessity of establishing the Daedeok Smart City Ecosystem*

The Daedeok Innopolis has problems, such as the insufficient collaboration and joint research among research institutes due to the aged

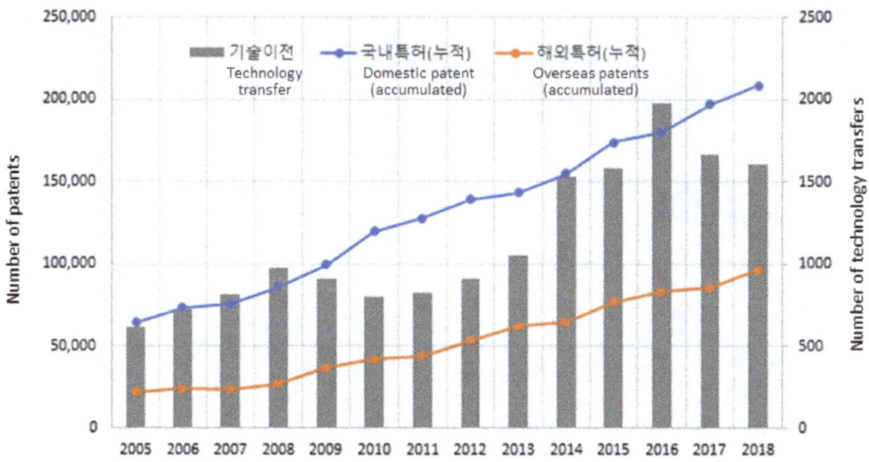

Figure 12.9. Status of patent and technology transfer of resident institutions in Daedeok Innopolis.
Source: Innopolis foundation, Statistical Data for R&D Special Zone.

infrastructure in the zone and the closed nature of government-funded research institutes, and the unsatisfactory ripple effect of R&D technology transfer and technology startups as for the R&D cost, while Daedeok Innopolis as Korea's leading basic research hub has achieved considerable results.

In particular, a concentration problem of population and economic power in the Seoul metropolitan area brought about the acceleration of concentration of innovative firms in the southern part of the capital [Kang, 2018] (see Figure 12.10). Apart from the capital area, only Daejeon Yuseog (Daedeok), Gwangju and part of Busan remain to be an area for growing innovative firms. Even regional innovative hubs in non-capital areas are under threat by the relocation of private R&D centers to the capital. The Hyundai Wia R&D Center was relocated in unity from Changwon to Uiwang in 2012, and the Doosan Heavy Industries & Construction Co. Ltd. Material Development Center was moved from Changwon to Yongin as well. The Samsung Heavy Industries R&D Center in Daedeok was also reduced with the construction of an R&D center in Pangyo, Seongnam. A newly-built LG Chemical Research Institute in Magok weakened its research function in Daedeok.

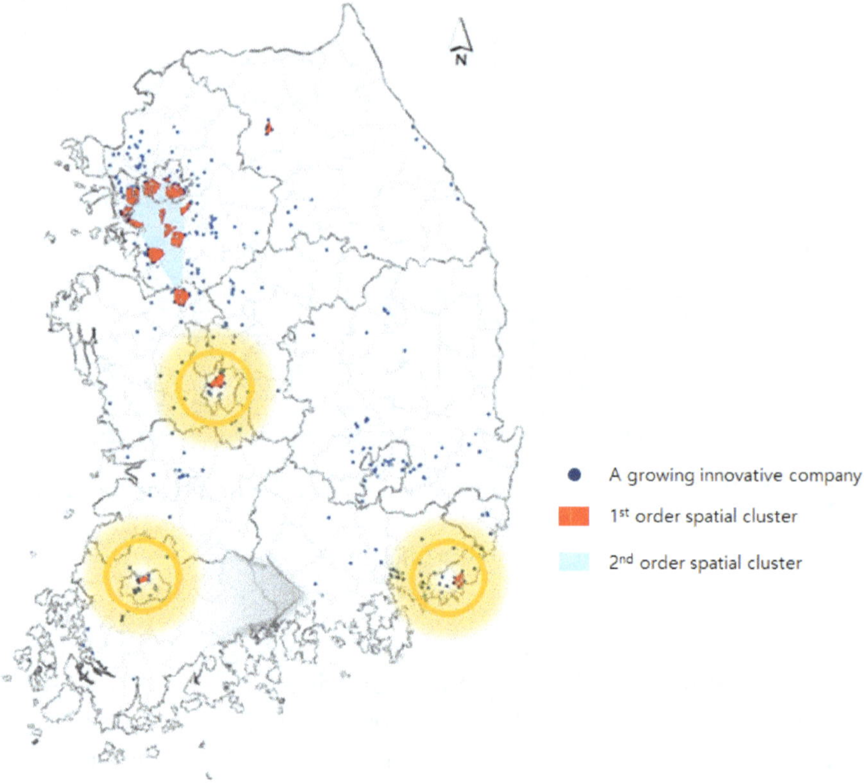

Figure 12.10. Distribution of innovative growth areas in Korea.
Source: Kang, 2018, p. iv.

Thus nowadays, the role of innovative cluster is not being fully played because industrial infrastructure like resident companies is relatively insufficient compared to research competence, even in Daedeok Innopolis, where innovative capabilities are comparatively abundant.

To solve this problem, current programs aim to make a breakthrough to strengthen the competitiveness of the Daedeok Innopolis by newly recreating space where companies can operate further in the existing integrated zone of research centers, and by establishing an ecosystem where commercialized technology, such as technology transfer, investment and market advancement, is well achieved, and by applying the concept of the smart city, which makes urban spaces as test beds for the development of new technology.

12.3.3 *Direction of building a Smart City Innovation Platform in the Daedeok Innopolis*

12.3.3.1 *Basic direction*

Building a Smart City Innovation Platform in the Daedeok Innopolis aims at R&D based on data generated in the urban space, and at a demonstration of its results there. At the same time, it aims to provide Living Lab where living convenience is provided to the local residents by making use of the latest technology and the locals give feedback on their experience about it with suggestions of new problems [Innopolis Foundation, 2020] (see Figure 12.11).

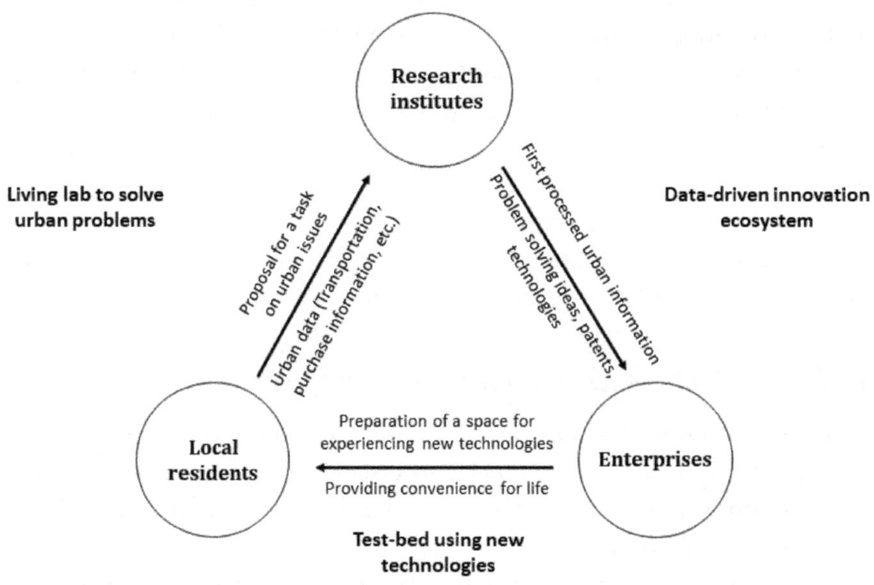

Figure 12.11. The concept of smart city innovation platform in Daedeok Innopolis.
Source: Innopolis foundation, 2020.

12.3.3.2 *Establishing an Innovation Ecosystem based on Data Industry*

For the innovative growth of Daejeon city, there must be disruptive innovation first in the Daedeok Innopolis, a source of growth. Above all, it is important to establish a data innovation ecosystem where the data produced in cities are freely shared and utilized.

This requires first, the promotion of free sharing and utilization of city-produced data by building an integrated platform for smart city management. For the construction of this platform, a systematic approach is needed including (1) designing a management system which collects, manages, and makes use of urban public data, (2) gradually applying it to hard infrastructure like transportation system, energy and water supply, and sewage and (3) expanding to soft infrastructure like education, welfare, medical services, and disaster prevention in the final.

Second, it is important to encourage the accumulation of public data and to expand users by disclosing the data of city and public institutes. An establishment of the public data set connected to traffic, parking, environment, crime prevention and street lamp system, etc., is needed in order to secure effective public data. In addition, building a system to collect and share domain data produced by industry activities, and vitalizing data community are important.

Third, it is also necessary to expand the base of data industry such as 5G network, IoT platform, and spatial information.

12.3.3.3 *Building test bed using new technology*

Unlike other cities, the Daedeok Innopolis concentrated with research centers and industrial firms needs to create an environment in which R&D output is boldly demonstrated and utilized to serve as test beds for advanced new technology.

To this end, it firstly needs to construct a test space for the operation and performance of new technologies and new products under a wide range of environmental conditions. Numbers of test spaces such as road type, residential type, park type and indoor type, would be provided in order for researchers and residents to experience new technologies and to give feedback on the results.

Second, it is necessary to build a platform to review the performance of new technologies and new products in a virtual space. The real-world data can be analyzed and simulated in the virtual space based on the same 3D modeling (digital twin), and its results can be fed back to solve the problem.

Third, a drastic regulatory innovation is needed for the application of innovative technologies. An introduction of a regulatory sandbox which removes regulations in a package only for smart cities is required in order to apply advanced technology and to provide various convergence-and-integration services. For this, there should be the vitalization of governance, involving number of entities and collaborative efforts of pan-governmental departments.

12.3.3.4 *Building living lab to solve urban problems*

In order to establish a smart city innovation ecosystem, it is important to convert the role of citizens, which has been simple consumers of urban services, to active participants in the entire process of urban construction management to work towards solving problems together. A circulation structure is required to propose creative ideas to solve urban problems and to give interactional feedback by an enterprise-business operation-government through technology testing and the development and discovery of services, etc.

The "Living Lab", an attempt to find solutions of social problems by using the field of life as a laboratory, can be a solution for urban problems. For the systematic advancement of the Living Lab, the following implementation plans are needed.

First, the construction and operation of an open platform where citizens propose ideas of problem-solving should proceed. Second, after selecting tasks to be performed from the proposed ideas, a pilot project should be implemented in order for the local government, research institutes and local communities to find a methodology of problem-solving.

Third, a collaborative project between research institutes and private companies which applies technologies and services to specific regions in the city is carried out to try to solve the problems. Fourth, local companies that have been transferred related technology are supported to secure competitiveness and to enter the market based on their track records gained from the concerned project.

The Ministry of Land has conducted a project inviting public participation in local governments nationwide to solve urban problems and to support the discovery of business models by making use of the smart city technology since 2019.

Accordingly, the city of Daejeon, where the Daedeok Innopolis is located, conducted five projects as Phase 1 around the Daejeon central open market last year, including (1) sharing parking lots, (2) prevention of electric fire, (3) safety net for drone (unmanned aerial vehicle, or UVA), (4) measuring network for fine dust density and (5) construction of data hubs.

It will be necessary to promote expert-challenging Living Lab in which research institutes and universities in the Daedeok Innopolis participate and discover solutions to the problems in the future.

12.3.3.5 *Establishing governance to construct Smart City Innovation Ecosystem*

Numerous innovative players including research institutes, universities, companies and residents participate in the Daedeok Smart City Innovation Platform. These innovative players are making their best efforts for the Daedeok Innopolis to take a leap into a global innovation cluster.

The Korea Innovation Foundation has its distinctions from other innovative entities in that it is a specialized institute to establish a governance for constructing the Smart City Innovation Ecosystem. It manages and operates the entire innovation ecosystem including collaboration and joint research between government-funded research institutes, technology transfer to private enterprises, startup business support, and fostering enterprises and collaborative development of industry and academy. Based on the technology competence and experience of technology commercialization accumulated in the Special Zone, a cooperative system is being established to plan and promote innovative startups with high quality and far-reaching influence while working with private experts like startup agencies. Additionally, numerous projects such as building a technology commercialization platform to effectively connect with public technology by finding out technology demand wanted by companies, and inviting investment to support enterprise growth and a program for advancing into the global market are being carried out.

12.4 Conclusion

The 4th Industrial Revolution era was first advocated at the World Economic Forum in early 2016, which is already treated as being past its prime by many people. However, the wave of change has just begun to have an influence on our lives. Nowadays, the startup business on the opposite side of the earth may ruin companies in our nations overnight, and a sudden withdrawal of global anchor enterprises in charge of the local economy might put an end to related companies. The global market is rapidly changing and integrated into one.

In order to overcome this crisis and to continue to develop the local economy, a technology commercialization ecosystem where value is created by transferring the technology of universities and research institutes to companies should be well built. This ecosystem works the most efficiently when the entities utilizing technologies such as companies, accelerators, and venture capitalists (VCs) are clustered around technology suppliers, including universities and research institutes. In this way, it is important to construct an innovation cluster around each local hub equipped with the infrastructure in order for innovative activities in the region to drive national innovative growth.

The concept of the smart city based on ICT technology might play an important role not only in strengthening the physical infrastructure of the innovation cluster but in establishing an innovation ecosystem as well. Overseas advanced cities are already using the smart city concept in many methods. Santander in Spain has been creating a Living Lab involving citizens and ICT companies based on the cooperation between public and private sectors, and making use of all the urban area as an experimental space for technology and service. By connecting about 20,000 sensors, collectors and cameras installed inside the city with the system, public institutes, general companies and citizens can utilize the data while taking notice of the urban situation in real time. Toronto in Canada is conducting numbers of business experiments (Cyber Physical Systems (CPS), autonomous public transportation, modular campus, etc.) [Ministry of Land, 2019] according to the characteristics of technology and project under the leadership of Google Sidewalk Lab.

Daedeok Innopolis must become a smart city linked to the innovative ecosystem beyond making the urban infrastructure smart. The construction of the innovation platform is required to research and develop a data-based innovative ecosystem based on the data generated in urban spaces, to demonstrate its achievements in the Daedeok Innopolis. This platform also provides local residents with living convenience by making use of the latest technology (i.e., new technology test beds), where they can also give feedback on their experiences about it and bring up new suggestions (Living Lab) as well.

At the same time, the urban environment where there is a melting pot of excellent and diverse cultures must be created by constructing the smart city infrastructure. Successful innovation clusters like Silicon Valley in the US and JungKwanChun in China are all closely connected to the industrial ecosystems in their regions by technology suppliers. These areas are full of vitality with young challengers in the startup business, as well as excellent universities and research centers, along with their art and culture in the cities. With frequency, creative young people who encounter superior technology create startups and employ talented personnel on investment. This is where innovative growth can happen.

The Daedeok Innopolis will be able to grow into a global innovation cluster if excellent technology and human resources drive startup businesses and its growth in the smart city, and an innovative ecosystem with new suggestions and feedback through citizens' participation is established.

References

IDC. (2019). *Market Prospect for Platform Business*. URL: https://www.idc.com/getdoc.jsp?containerId=prAP46180120.

Innopolis Foundation. (2019). *A Study on the Implementation of Space Activation for the Master Plan for Renovation of Daedeok R&D Special Zone*. Innopolis Foundation.

Innopolis Foundation. (2019) *Statistical Data for Special Zone*. URL: https://www.innopolis.or.kr/board?menuId=MENU00366&siteId=null.

Kang, H.J. (2018). *Research on Location Policy for Innovative Enterprises in the 4th Industrial Revolution Era*. Korea Research Institute for Human Settlements.

Kim, S.B. (2017). *The 4th Industrial Revolution Era and Smart Regional Innovation*. Korea Institute for Industrial Economics and Trade.

Ministry of Land, Infrastructure, and Transport. (2019). *The 3rd Comprehensive Plan for Smart City 2019–2023*.

Ministry of Science and ICT. (2018). *The 3rd Comprehensive Plan for R&D Special Zone Development* (revised).

Park, Y.S., and Glenn, J. (2019). *World Future Report 2020*. Business Books.

Samjong KPMG Economy Research Institute. (2019). The successful strategy for the platform business, *Samjong Insight*, 67. URL: https://assets.kpmg/content/dam/kpmg/kr/pdf/2019/kr-insight67-platform-20191121.pdf.

The 4th Industrial Revolution Committee. (2018). *Promotional Strategy for Smart City for Urban Innovation and the Creation of Future Growth Engine*. Ministry of Land Transportation, Korea. URL: https://www.korea.kr/archive/expDocView.do?docId=38138.

© 2023 World Scientific Publishing Company
https://doi.org/10.1142/9789811257186_0013

CHAPTER 13

Smart City Energy Inclusion, Towards Becoming a Better Place to Live

Sang-Ryong Cha

Department of Public Policy, Regional Design and Development, University of Nagasaki, Sasebo, Nagasaki, Japan
Email: cha@sun.ac.jp

13.1 Introduction: Where Is a Better Place to Live?

Where is a better place to live? Regarding this question, since 2007, the Economist Intelligence Unit (EIU) has presented its annual ranking of the world's most livable cities with respect to urban quality of life based on assessments of stability, healthcare, culture and environment, education, and infrastructure, as displayed in Figure 13.1, which shows changes in scores over a 10-year period.

Besides this, numerous measurement tools ranking cities in terms of livability have been developed. These include some famous indices, for instance, the Mercer Quality of Living Index and the OECD Better Life Index. Although all of the indices consist of similar indicators and categories, each index places significant importance on a specific category of livability, for example, health in the EIU and Mercer indices, and the economy in the OECD index, as presented in Table 13.1. This highlights

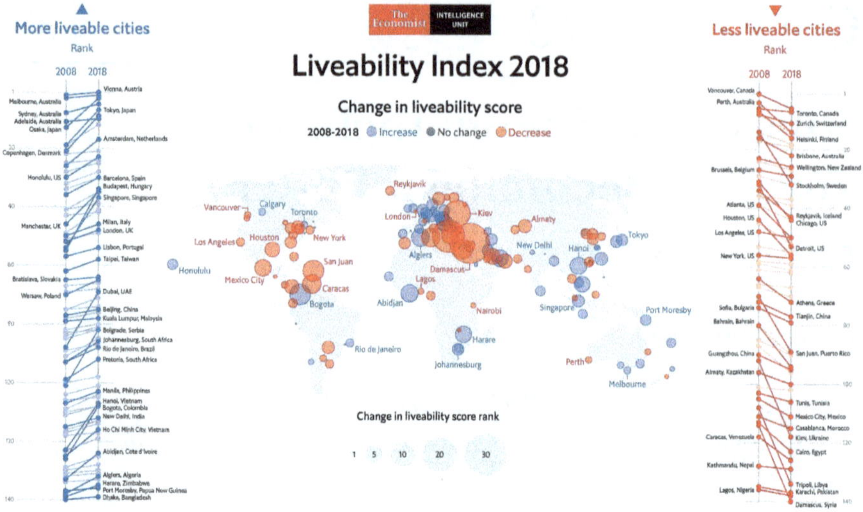

Figure 13.1. Changes in livability scores from 2008 to 2018.
Source: From *The Economist*, https://economist.com, [August 16, 2018].

the fact that the citizens of a society pursue suitability and satisfaction simultaneously in determining a livable place.

As globalization proceeds, the world becomes more diversified. Although it has been understood that diversity is a source of innovation and revolution, it is difficult to satisfy increasingly diversified demands and suit countless new situations in an ever-changing and uncertain world. Moreover, even the natural environment is changing because of global warming. Therefore, in the coming era when change is the only constant [Harari, 2018], a better place to live is not simply a livable place but also an adaptable place with a flexible system able to suit any new situation in and about the place in terms of diversity/diversification.

The ability to deal with diversity, to learn new situations, and to preserve our living balance in unfamiliar situations will be of great importance. In order to be prepared for the era to come, reinventing our whereabouts and adapting to changing situations is necessary. However, dealing with diversification can be challenging because, in general, it is invariably stressful. In fact, many people may simply dislike

Table 13.1. Categories and indices of three major livability indexes.

EIU index	
Categories	**Indices**
Stability	Prevalence of petty crime Prevalence of violent crime Threat of terror Threat of military conflict Threat of civil unrest/conflict
Healthcare	Availability of private healthcare Quality of private healthcare Availability of public healthcare Quality of public healthcare Availability of over-the-counter drugs General healthcare indicators
Culture and Environment	Humidity/temperature rating Discomfort of climate to travelers Level of corruption Social or religious restrictions Level of censorship Sporting availability Cultural availability Food and drink Consumer goods and services
Education	Availability of private education Quality of private education Public education indicators adapted from World Bank
Infrastructure	Quality of road network Quality of public transport Quality of international links Availability of good quality housing Quality of energy provision Quality of water provision Quality of telecommunications

Mercer quality of living index	
Categories	**Indices**
Political & Social Environment	Relationship with other countries, Internal stability, Crime, Law enforcement, Ease of Entry and Exit
Medical and Health	Hospital services, Medical supplies, Infectious diseases, Water potability, Sewage, Waste removal, Air pollution, Troublesome and destructive animals and insects
Public Services and Transport	Electricity, Water availability, Telephone, Mail, Public transport, Traffic congestion, Airport
Consumer Goods	Meat and fish, Fruits and vegetables, Daily consumption items, Alcoholic beverages, Automobiles
Economic Environment	Currency exchange regulations Banking services
Socio-cultural Environment	Limitation on Personal Freedom Media and censorship
School & Education	Schools
Recreation	Variety of restaurants, Theatrical and musical performances, Cinemas, Sport and leisure activities
Housing	Housing Household appliances and furniture Household maintenance and repair
Natural Environment	Climate Record of natural disasters

OECD BLI	
Categories	**Indices**
Housing	Dwellings without basic facilities Housing expenditure Rooms per person
Income	Household net adjusted disposable income Household net financial wealth
Jobs	Labor market insecurity Employment rate Long-term unemployment rate Personal earnings
Community	Quality of support network
Education	Educational attainment Student skills Years in education
Environment	Air pollution Water quality
Civil Engagement	Stakeholder engagement for developing regulations Voter turnout
Health	Life expectancy Self-reported Health
Life	Life satisfaction
Safety	Waking during night Homicide rate
Work-life Balance	Employees working very long hours Time devoted leisure and personal care

Source: Randhawa and Kumar [2017, pp. 281–282].

adapting to new situations caused by diversification, especially as they age. If individuals lack the ability to adapt, it is difficult for them to accept something new, and it may become tempting for such people to neglect or exclude diversity from their living places. This indicates that people try to be content with habituated present conditions, meaning that chances for innovation and revolution ultimately leading to a better life may be lost.

Can technology help with acquiring an ability to adapt? Indubitably, technology can be of great help; however, what specific technology is required is the main point to consider. Regarding this, customization and real-time operation are necessary to realize an adaptable place for a better life technologically because the ability to deal with diversity is simply the ability to adapt to more individualized needs caused by the diversified world. Customization aims to accurately meet needs; therefore, the collection, integration, and analysis of Big Data from each sensor making the best use of artificial intelligence (AI) is necessary. On the other hand, real-time operation aims to manage needs immediately without a time lag; thus, obtaining more lightweight operating system (OSs) for the Internet of Things (IoT) is required. Both are primary technological characteristics of a flexible system for an adaptable place. Smart cities can be expected to employ a flexible system applying technologies of self-monitoring and self-response, thereby acting as promising city models towards becoming better places to live.

13.2 Hitherto: Issues Concerning Energy Demand and Supply

At present, it may be said that the essence of a smart city is technology because it has never been emphasized like this in all directions in any new city models. If what is wanted for a better place to live is understood, smart city technology can assist with realizing this; however, if what is wanted is unknown, it will be too straightforward for a smart city to shape lifestyles and take control of life. More specifically, as a smart city improves its understanding of its citizens, people might increasingly find themselves serving it, instead of it serving them. Therefore, to succeed in the development of smart cities, what is expected of the flexible system of a smart city must be understood.

Energy demand and supply is a crucial issue concerning the expectations of a flexible system of a smart city because it is indispensable to everyday living, especially city living. For example, it is the civilian sector, consisting of the commercial sector and the residential sector that has influenced energy consumption in Japan since the 1970s due to urbanization and the expansion of the middle class, which has consumer spending power. That is, the formation of an industrial society brought about rapid urbanization and an emerging middle class, and both stimulated energy consumption in the civilian sector. Moreover, energy consumption regarding passengers in the transportation sector has risen because of developments in public transportation and the boom in private vehicle ownership, which is closely connected with urban lifestyles. Accordingly, a relationship between energy consumption and the quality of life in cities can be seen, as displayed in Figure 13.2.

To address the rise in the demand for energy, the use of high density energy including coal, oil, and nuclear energy has been extended, and energy diversification, for example, using a variety of energy sources in order to secure the stability of energy supplies, has shown a tendency to be conspicuous since the oil crises in the 1970s. In particular, since the Great East Japan Earthquake and the Fukushima nuclear incident in 2011, it appears that the energy diversification of power supplies has been a matter of overriding concern regarding the energy demand and supply issue based on the understanding of the relationship between energy consumption and the quality of life in cities. In Figure 13.3, an overview of the history of energy in Japan can be seen.

However, the main concern is location, especially the scale of place to build grids to achieve a sustainable urban energy system. Scale with respect to the energy balance between supply and demand needs to be considered. The origin of energy resources and where conversion takes place must also be considered. For instance, regarding Japan's energy balance flow in FY2017, there was a loss of approximately 30% resulting from transmission or distribution loss and self-consumption of the power/conversion sector during the conversion of energy between the

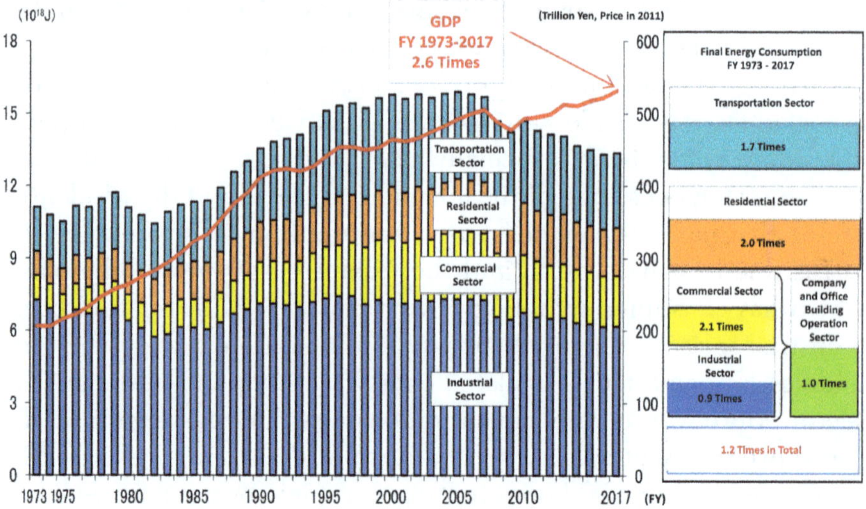

Figure 13.2. The transition of energy consumption and real GDP in Japan.
Source: The Agency for Natural Resources and Energy [2019, p. 104].

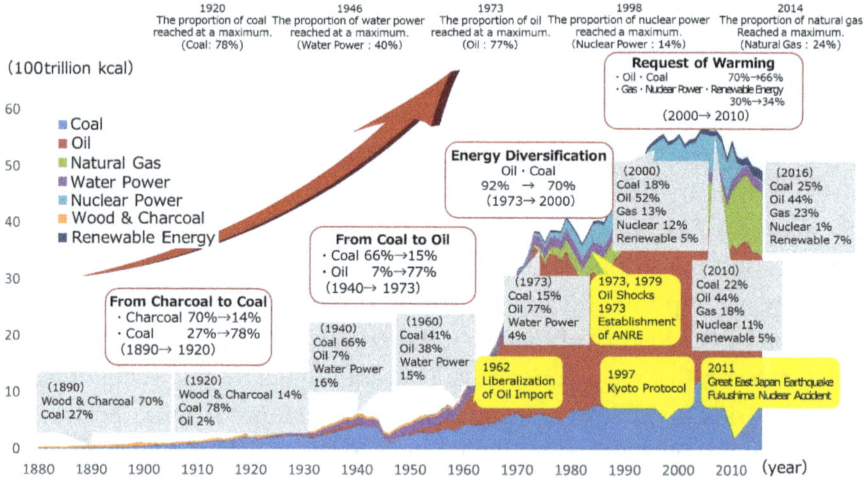

Figure 13.3. The history of energy in Japan: Changes in the structure of energy demand and supply.
Source: The Agency for Natural Resources and Energy [2018, p. 8].

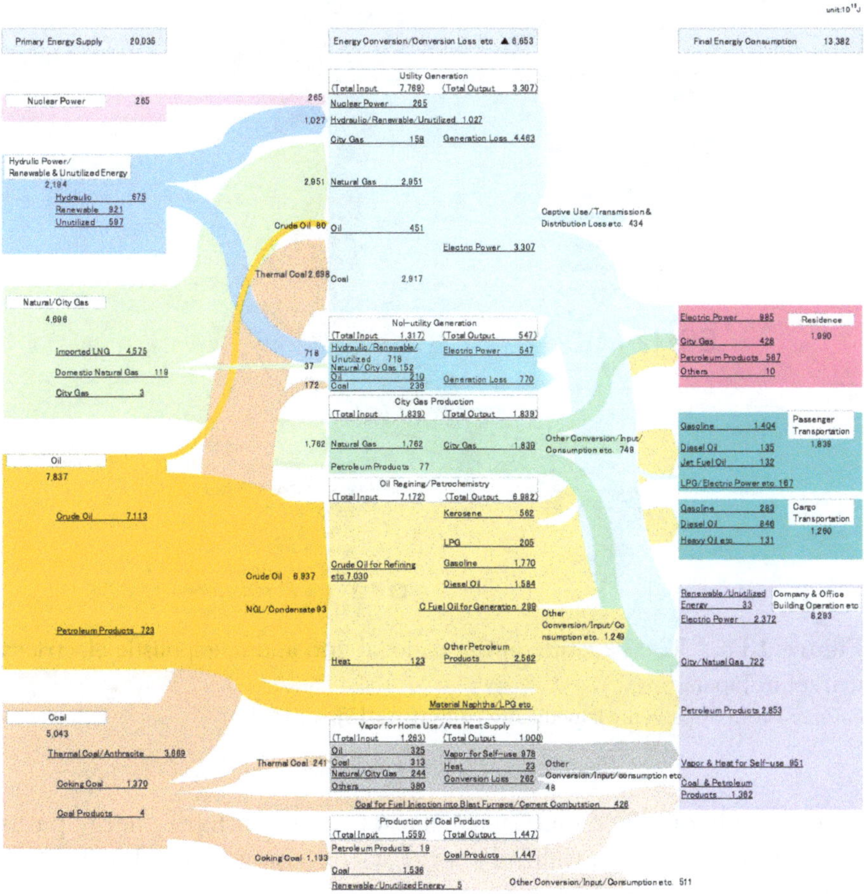

Figure 13.4. Energy balance flow in Japan (FY2017).
Source: The Agency for Natural Resources and Energy [2019, p. 10].

primary energy supply and final energy consumption. In Figure 13.4, the energy balance flow in Japan can be seen.

Above all, in the electricity sector, loss is generally caused by the structure of the national grid and the market. Unlike most other developed countries, the national grid of Japan is divided into eastern and western girds, and 10 regulated major power companies dominate the

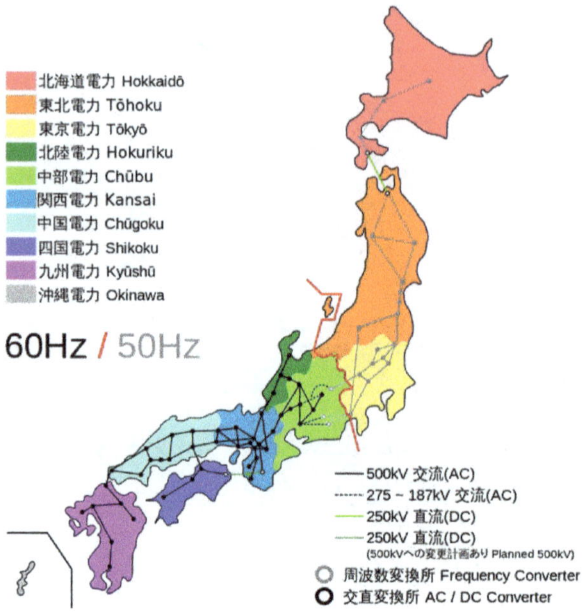

Figure 13.5. The structure of the national grid and oligopolistic electricity market in Japan.
Source: From https://en.wikipedia.org [July 12, 2018].

domestic oligopolistic electricity market based on hard energy paths supported by huge centralized energy systems on a national scale. This is depicted in Figure 13.5.

Besides loss, the traditional hard energy paths mentioned earlier based on fossil and nuclear fuels have already faced difficult problems in terms of energy flexibility or resilience, in particular. Following the Fukushima nuclear accident in 2011, a trend toward alternative soft energy paths [Lovins, 1977] supported by small diversified energy systems on a local level has intensified with the goal of building a nuclear-free society in Japan. However, a soft energy path based on renewable energy sources has some limitations regarding stability, variability, and supply potential in spite of the low economic/technological burden acquired to realize it; see Figure 13.6 for a representation of the differences between the two energy paths.

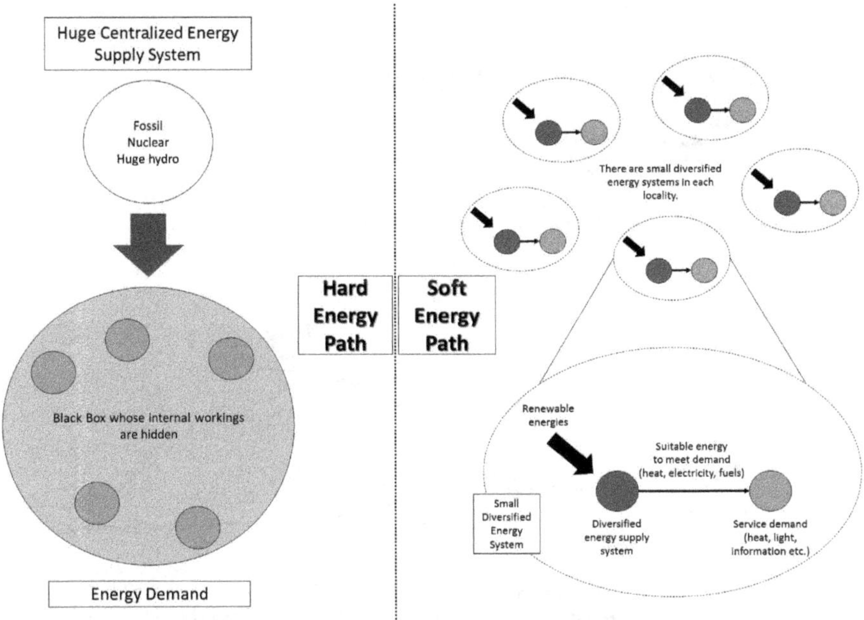

Figure 13.6. Conceptual differences between two energy paths: Hard and soft. *Source*: Shimoda [2014, pp. 17, 19].

It seems that further reconsideration of scale of place is necessary in order to minimize the loss rate of energy conversion if improving the energy efficiency of present energy systems is a goal. From this point of view, a holonic path can be adopted [Kaya *et al.*, 1988] as an alternative, incorporating soft energy paths with hard energy paths complimentarily based on load management. Figure 13.7 demonstrates the features of such a path.

In the concept of a holonic path, load management is a way of raising the availability of supply facilities by regulating demand patterns using an information system to inform customers of suppliers' situations. This is based on the mutual cooperation between suppliers and customers and is the same as the meaning of energy management in the context of a smart city. That is, the energy system of a holonic path has a high affinity with the flexible system of a smart city.

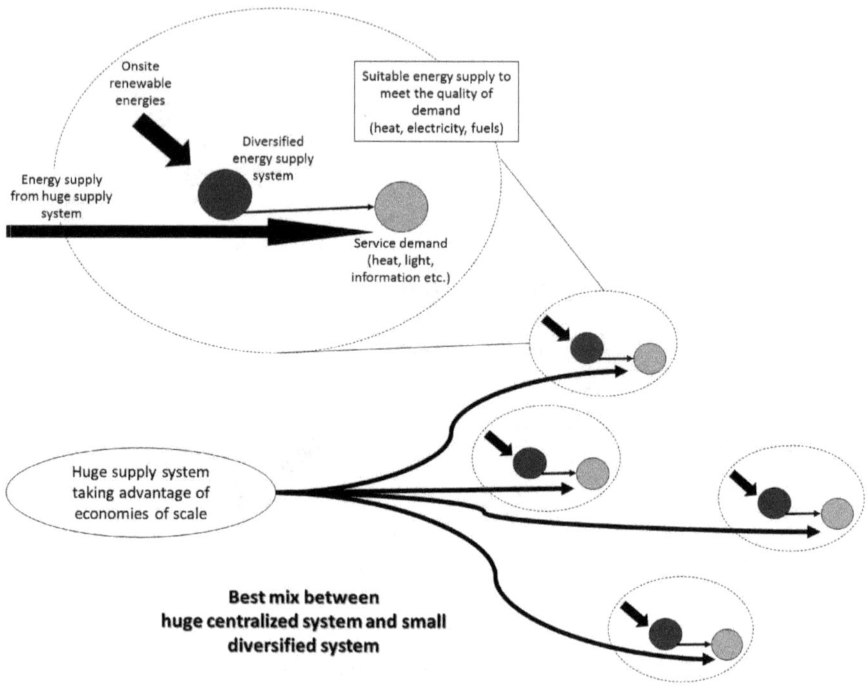

Figure 13.7. Conceptual features of a holonic path.
Source: Shimoda [2014, p. 21].

13.3 Hereafter: Issues Concerning Energy Inclusion

At a system level, the purpose of all of the paths mentioned earlier is not energy itself but the service it provides. Although the synergistic relationship between energy consumption and the quality of life in cities is already understood, energy services actually expected in city life are fixed to some extent. If an expected energy service is fixed, the conclusive factor in choosing a more appropriate system for a better place to live is accessibility to energy services because an expansion in the supply of energy and a reduction in the demand for energy are identical with respect to energy management, which, as stated by Lovins [1985], is the very concept of negawatt power. Negawatt power seems to emphasize the effectiveness of energy conservation; however, the vital point of the concept is the affordability of energy services in terms of customer needs. That is, customer actions for energy conservation based on the negawatt power concept are directly rewarded by the savings on charges. It becomes a cost incentive to encourage more customers to engage in the negawatt movement called 'the power-saving plant', or *Setsudensho* in Japanese, in Japan.

The power-saving plant is based on the concept of a minimum cost scheme that requires supply companies to implement all customer actions for energy conservation cheaper than the energy supply cost prior to supply expansion [Shimoda, 2014]. Thus far, energy conservation investment remains in the first and the second steps only, but if scale of place to put an energy system into practice with relatively small investment for energy conservation higher than for energy supply can be found, energy conservation by the power-saving plant can progress to the third step at a system level. This concept is presented in Figure 13.8.

The power-saving plant at a system level involves various technologies and institutions reducing energy consumption in society as a whole by using electricity more efficiently. This means that the development of the power-saving plant can be a catalyst for forming a quadruple innovation helix as a synthesis of top-down policies and practices from the government, universities, and industry, balanced and shaped by bottom-up initiatives and actions by civil society [Monteiro and Carayannis,

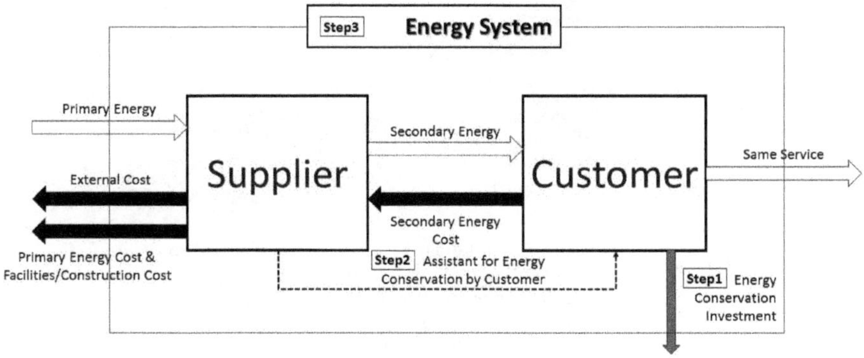

Figure 13.8. Minimum cost scheme.
Source: Shimoda [2014, p. 25].

2017]. Actually, in Kitakyushu, a model city for green growth chosen by the OECD [OECD, 2013], the Kitakyushu smart community centered around the community energy management system (CEMS) as a local power-saving plant has been developed at the site of the Nippon Steel Corporation Yahata Iron Factory by a Quadruple Helix stakeholder collaboration consisting of the City of Kitakyushu (municipality), Nippon Steel Corporation, Kyushu University, and residents from approximately 200 households, etc., since 2011, as depicted in Figure 13.9.

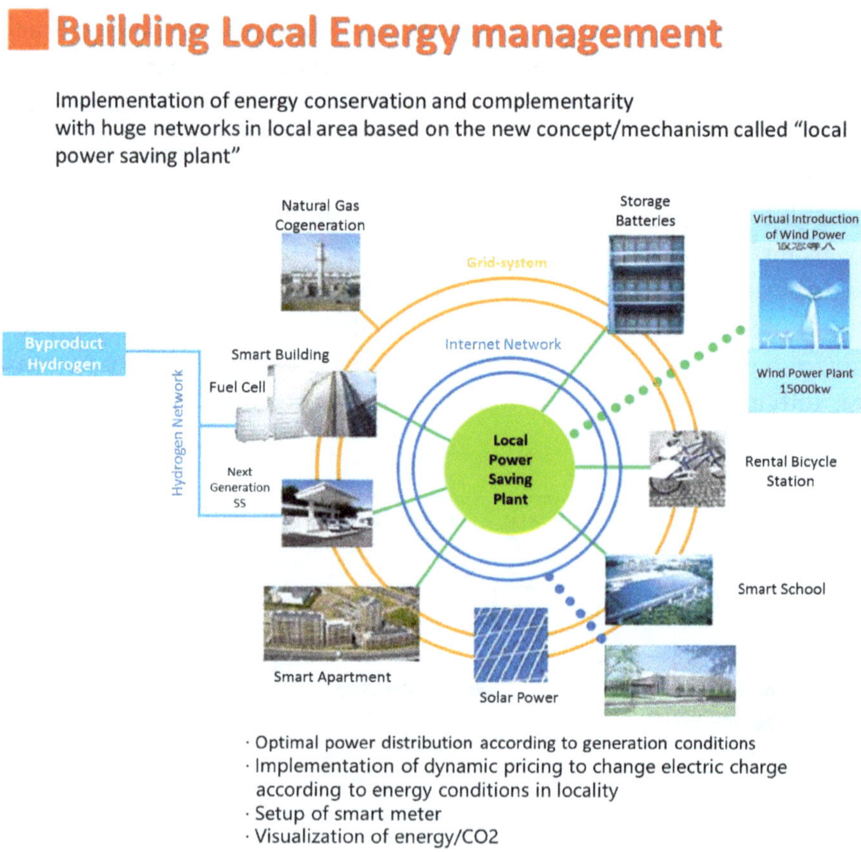

Figure 13.9. A conceptual sketch of a local power-saving plant in Kitakyushu. *Source*: From City of Kitakyushu, http://www.city.kitakyushu.lg.jp [December 12, 2011].

In this smart community, customers have the opportunity to be prosumers [Toffler, 1980], not only of the energy service itself but also of the energy system, utilizing tools for open innovation, for example, urban living labs [Marvin *et al.*, 2018] or open data hackathons/ideathons such as the Bristol case in the UK [Kohno, 2017]. In a territorial context, they can recognize and define energy issues, find appropriate directions for solutions, and decide the priorities of energy services. Consequently, energy services can develop into more practicable and serviceable services for most local customers, and this means that the reliability of energy services is improved.

Affordability and reliability constitute the accessibility of energy services. The former encompasses the level of energy self-sufficiency. The latter encompasses the extent of energy saving. From this point of view, it seems that small diversified soft energy path systems have a clear advantage over huge centralized hard energy path systems. However, some insuperable limitations remain, so it is reasonable to state that both energy systems should continue to coexist in multiplexing energy systems, adopting a holonic path to establish and maintain reliable and affordable access to energy services covering households' or enterprises'

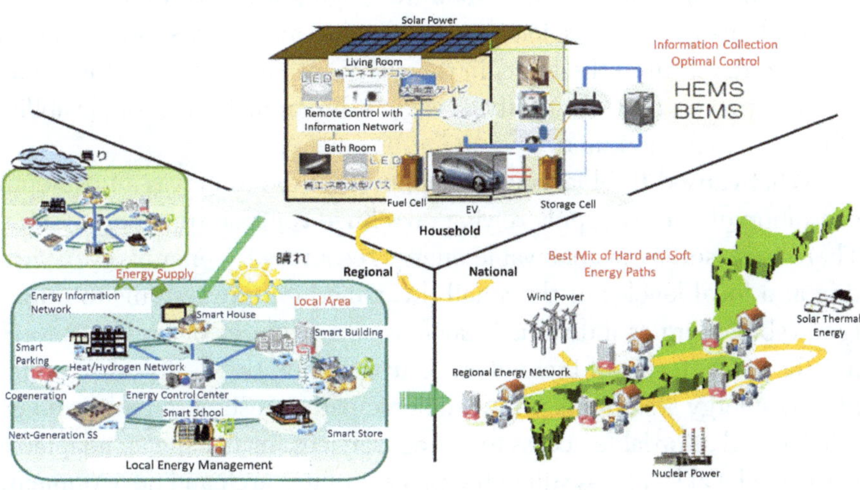

Figure 13.10. The Japanese version of a smart grid.
Source: The Agency for Natural Resources and Energy [2010, p. 149].

basic energy needs. If such a system is embodied in a smart city concept, it is nothing else but smart energy inclusion.

Actually, since 2010, this has been embodied in the Japanese version of a smart grid initiated by the Agency for Natural Resources and Energy, and the regulation of supply and demand, known as load management, has been implemented as a demand response to induce the fluctuation of demand by announcing changes in the price of electricity in response to suppliers' situations. The Japanese version of a smart grid is presented in Figure 13.10.

13.4 Conclusion: How a Better Place to Live in the Era to Come Be Built?

On 6 September 2018, all of the citizens in Sapporo, the largest city on Hokkaido Island, experienced the first major blackout in Japan because the Tomatoatusma Coal Power Plant, the biggest power plant in the region, went offline due to the 2018 Hokkaido Eastern Iburi Earthquake. Although Sapporo is considered a greatly livable city in Japan, this city was completely unable to devise any appropriate measures to manage the situation when the huge power plant covering half of the demand for electricity on Hokkaido came to a sudden stop in only two days. All of the lights in the city went out, and all of the urban facilities powered by electricity ceased to function. That is, the livability of the city malfunctioned.

What caused the blackout in Sapporo? The reason was an overconcentration of energy supply relying on a huge centralized energy system. There were sufficient renewable energy resources for energy diversification at a local level, but the small diversified energy systems to utilize renewable energies individually such as smart houses or buildings were insufficient in terms of technology. Furthermore, the flexible and multiplexing energy system adopting a holonic path to establish and maintain reliable and affordable access to energy services for making an adaptable place at a local level was still in the process of research and development.

The Government of Japan proclaimed that renewable energy must become the main source of power in Japan by 2050 in its Fifth Strategic

Energy Plan. In response, Sony has vowed that all of the energy it uses will originate from renewable sources by 2040. In this situation, it is expected that smart energy inclusion will be realized by the Japanese version of a smart grid based on the concept of a minimum cost scheme and demand response for making a better place to live in the near future.

References

Harari, Y.N. (August 12, 2018). Yuval Noah Harari on what the year 2050 has in store for humankind. URL: https://www.wired.co.ku.

Kaya, Y., Nakagami, H., Suzuki, H., and Nishihiro, D. (1988). *New Era of Energy: Towards a Holonic Path* (in Japanese). Tokyo: The Energy Conservation Center, Japan.

Kohno, M. (2017). *Latest landscape of smart cities in the West 2017: An Observation Through the Eyes of a Japanese Smart City Visionary*. Tokyo: Michi Creative City Designers Inc.

Lovins, A.B. (1977). *Soft energy paths: Towards a Durable Peace*, p. 231, Harmondsworth; New York [etc.]: Penguin.

Lovins, A.B. (1985). Saving Gigabucks with Negawatts, *Public Utilities Fortnightly*, *115*(6), 19–32.

Marvin, S., Bulkeley, H., Mai, L., McCormick, K., and Palgan, Y.V. (2018). *Urban Living Labs: Experimenting with City Futures*. Routledge.

Monteiro, S.P.D.O., and Carayannis, E.G. (2017). *The Quadruple Innovation Helix Nexus: A Smart Growth Model, Quantitative Empirical Validation and Operationalization for OECD Countries*. New York: Palgrave Macmillan.

OECD. (2013). *Green Growth in Kitakyushu, Japan* (OECD Green Growth Studies), p. 127, OECD Publishing.

Randhawa, A., and Kumar, A. (2017). Exploring Livability as a dimension of Smart City Mission (India), *International Research Journal of Engineering and Technology (IRJET)*, *04*(11), 277–285.

Shimoda, Y. (2014). *Introduction to Urban Energy System: Housing, Architecture, Urban Energy Saving, and Low Carbonization* (in Japanese), p. 263. Kyoto: Gakugei Shuppan Sha.

The Agency for Natural Resources and Energy. (2010). *Japan's Energy White Paper 2010* (in Japanese), Ministry of Economy, Trade and Industry.

The Agency for Natural Resources and Energy. (2017). *Japan's Energy White Paper 2017* (in Japanese), Ministry of Economy, Trade and Industry.

The Agency for Natural Resources and Energy. (2018). *Japan's Energy White Paper 2018* (in Japanese), Ministry of Economy, Trade and Industry.

The Agency for Natural Resources and Energy. (2019). *Japan's Energy White Paper 2019* (in Japanese), Ministry of Economy, Trade and Industry.

The Data Team. (August 16, 2017). Global liveability has improved for the first time in a decade—Daily chart. URL: https://www.economist.com.

The Environment Bureau. (December 12, 2011). Building Local Energy Management System. URL: http://www.city.kitakyushu.lg.jp.

Toffler, A. (1980). *The Third Wave: The Classic Study of Tomorrow*, p. 560. Bantam Books.

Wikipedia. (July 12, 2018). Electricity sector in Japan. URL: https://en.wikipedia.org/wiki/Electricity_sector_in_Japan.

© 2023 World Scientific Publishing Company
https://doi.org/10.1142/9789811257186_0014

CHAPTER 14

Applications of a Smart City for Sustainable Urban Regeneration

Sunghee Lee,[*] Jongkon Kim,[†] Youngchul Kim[‡,§]

[*]*KAIST Urban Design Lab, Department of Civil and Environmental Engineering, Korea Advanced Institute of Science and Technology, Daejeon, Republic of Korea*
[†]*Smart City Business Group, Seoul Housing & Communities Corporation, Seoul, Republic of Korea*
[‡]*KAIST Urban Design Lab, Department of Civil and Environmental Engineering, Korea Advanced Institute of Science and Technology, Daejeon, Republic of Korea*
[§]*Corresponding author, Email: youngchulkim@kaist.ac.kr*

14.1 Introduction

Many cities around the world have experienced rapid population growth and urbanization since the Industrial Revolution. In continuing urban population growth and urbanization, cities have caused environmental, economic and social problems, such as environmental pollution due to overcrowding and regional decline due to the sudden loss of central functions (e.g., unemployment rate, poverty, etc.) [Bolici and Mora, 2015; Chourabi *et al.*, 2012; Yigitcanlar, 2009]. In addition, the quality of life of urban residents has significantly deteriorated [Bolici and Mora, 2015] because other urban problems such as aging, a low birthrate, and an abnormal climate are mutually influenced and have expanded to larger urban problems [Shen *et al.*, 2012]. According to a United Nations (UN) report [2018], 55% of the world's population currently lives in urban areas, and 68% of the world's population will live in urban areas by 2050. Therefore, this phenomenon is expected to continue in the future.

Cities have developed various urban planning strategies to ensure urban sustainability by solving urban problems and responding to future needs. Recently, advanced information and communication technology (ICT) and new technology-based solutions have been recognized as important means to enable sustainable cities with the wave of the Fourth Industrial Revolution, and the concept of smart cities is drawing attention [Bibri, 2019; Sharifi, 2020]. Smart technologies and smart cities that are well adjusted with the goal of sustainability are likely to bring about unprecedented major changes in various areas of cities and advance in the long term. Therefore, leading smart city planning can determine the competitiveness of cities and can act as a driving force for the growth of a country. Accordingly, smart cities are receiving great attention from decision makers and policymakers [Bakıcı *et al.*, 2013; Bibri, 2019], and various plans and projects are in progress globally [Sharifi, 2019].

Smart cities aim to improve the quality of life of urban residents and systematic management through efficient resource allocation and use, convenient service provision, and solving numerous urban problems. Smart cities also have the potential to comprehensively handle urban components, functions, policies, and systems. Based on the goals and potential of smart cities, if smart cities are combined with regeneration strategies and plans in regions that require new and complex regeneration due to physical obsolescence or changes in social and economic environments, more effective urban regeneration and transition to a smart city will easily occur. The scope of urban regeneration has reached the level of solving a wide range of social, cultural and economic problems, beyond simply improving the aging physical environment [Couch *et al.*, 2011]. In addition, the perspective of urban regeneration, which aims to lead in a sustainable direction by managing the local resources and population, is quite consistent with that of a smart city. Above all, the grafting of smart cities to urban regeneration can control the indiscriminate application of smart technologies or solutions, since there is a clear purpose and specific requirements for urban regeneration, and achieve urban regeneration in a more rational and innovative way than before.

In this chapter, we propose the direction of smart city planning as a new paradigm for urban regeneration and introduce some application cases. In detail, we will examine the meaning and basic direction of smart cities from the perspective of urban planning and identify the dimensions and requirements essential for smart urban regeneration. Then, we will derive implications by introducing applied cases for each essential dimension.

14.2 Smart City and Urban Regeneration

14.2.1 *Concept of a smart city*

Smart cities began to emerge with the transition to the Fourth Industrial Revolution. However, the current concept of smart cities spread after the late 2000s, and is the result of continuous development for approximately 15 years to the present [Sharifi, 2019]. Prior to the use of the term "smart city", there were the "digital city", "intelligent city", "ubiquitous city", "hybrid city", and "knowledge city" concepts [Albino *et al.*, 2015; Anthopoulos and Fitsilis, 2010; Sharifi, 2020; Yigitcanlar *et al.*, 2008]. Although the meaning and direction of these concepts are slightly different, it can be seen that new proposals for urban management have been continued based on innovation through developing ICT and knowledge [Angelidou, 2015; Viale Pereira *et al.*, 2017]. "Smart city" has been developed into a more innovative concept while encompassing the previous concepts. However, there is no consensus on the definition of a smart city, and a smart city is defined differently according to the direction pursued by each country or local government, institutions and organizations [Bibri, 2019; Nam and Pardo, 2011]. Smart cities are the result of the combination of urban planning aimed at urban sustainability and resilience through technological progress leading from the Third Industrial Revolution to the Fourth Industrial Revolution and the connection of people–information–knowledge–city. Therefore, a smart city means a city that can efficiently operate a city and provide high-quality services to users by achieving optimization and advancement of the urban environment through the fusion of technology and knowledge.

Smart cities can comprehensively address the goals of cities such as the resource and energy allocation, climate stability, disaster management, public safety, and welfare (inequality, insecurity, aging, and employment) within cities. In addition, smart cities can encompass a wide range of strategies and plans for improving the quality of life in cities, providing high-quality public services, and managing efficient cities. For the successful realization of this plan, the establishment of an infrastructure and a platform for data collection, the composition of governance and citizen participation, and the creation of an environment for innovation are essential.

First, data collection, analysis, and monitoring in smart cities are key sources for urban operation and management and efficient resource allocation. Therefore, it is essential to establish a physical technology infrastructure for collecting data in cities and a virtual platform that can operate the data. Cities need to improve their infrastructure and spatial structure so that the concept can be flexibly applied. In addition, it should be possible to control and manage all flows of a smart city through the establishment of a data center or an integrated control center (i.e., a physical hub that can integrate, process, and analyze various city data).

Second, the active participation of various actors is the basis of quickly and stably converting a city to a smart city. In particular, it is necessary to actively engage citizens, the main body of smart cities. Citizens' participation plays a very important role in several stages, such as data collection and feedback processes for operating smart cities. Furthermore, the more participation is strengthened, the more user-oriented urban planning can be realized. In addition, harmony between the public and private sectors and collaboration with expert groups in each field are essential. Therefore, it is necessary to establish and operate governance composed of various actors and to strengthen the connection and cooperation between them.

Third, creating an environment for innovation is an essential aspect that can influence the sustainability of smart cities. In smart cities, innovation is the driving force leading smart cities. Therefore, for the innovation of smart cities, it is necessary to establish a cluster and create a small-scale activity space along with a strong strategy for policy and institutional support. Cities should provide a foundation for attracting or fostering innovative companies; encouraging start-ups; and establishing a cooperative system between universities, research institutes, companies and industries. Based on this, advanced smart technologies, products and industries must be constantly created. Furthermore, user participation is indispensable in order to improve the service efficiency in a city and continuously improve it. Furthermore, online and offline platforms should be provided to communicate and realize ideas.

14.2.2 *Smart urban regeneration: Smart city as a new paradigm for urban regeneration*

Urban regeneration aims to (1) regain urban competitiveness, (2) achieve housing welfare and social inclusion and (3) revitalize the economy for cities that are experiencing indiscriminate expansion, changes in population and industrial structure, and physical decline. Therefore, urban regeneration focuses on improving the physical environment, strengthening the capacity of residents, and reallocating and using local resources. In this process, the introduction of the smart city concept can be an effective means. This is because smart cities basically collect and analyze citywide information, plan customized strategies to solve each city's conditions and problems, and induce the efficient allocation and use of resources to solve problems more effectively than conventional methods. In addition, smart cities create an open environment in which citizens themselves can recognize their urban situation and participate directly, thereby enhancing citizens' perception and satisfaction. This can bring about effects such as improvement of city services, quality of life, and cost reduction. In addition, smart cities are expected to enhance the sustainability and resilience of cities through improvements in each sector such as transportation systems, water resources, energy and waste, and the environment.

For successful smart urban regeneration, 'smart environment', 'smart participation', and 'smart space' are required in Figure 14.1. The 'smart environment' refers to the physical environment and services required at each stage in the entire process of achieving smart urban regeneration. This is the basic foundation for helping various smart city

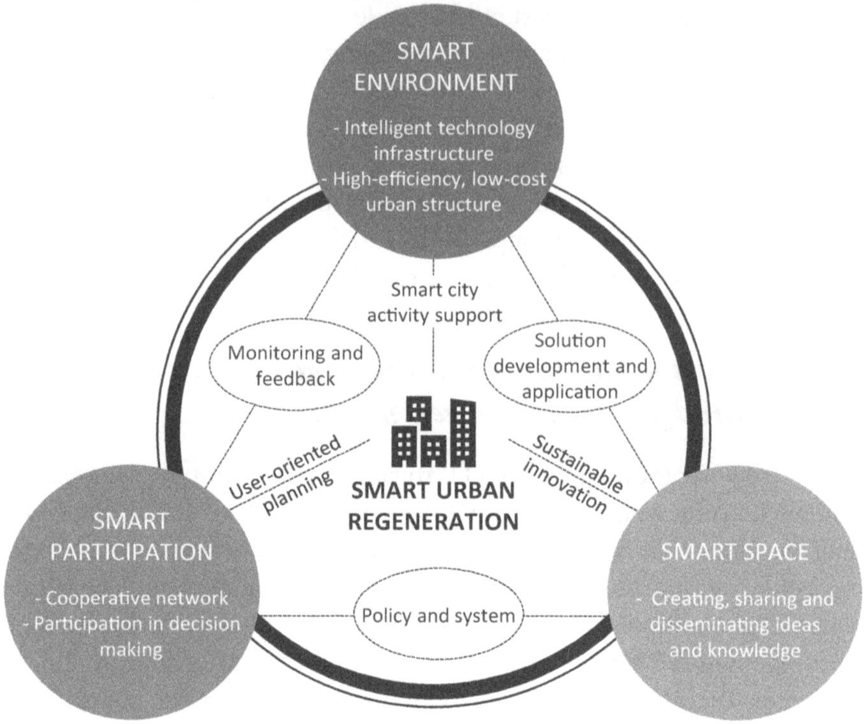

Figure 14.1. Essential requirements for smart urban regeneration.

activities to be conducted smoothly, and it must be supported so that solutions can be advanced and maintained in the future. A smart environment aims at the transition to an urban spatial structure that can increase the efficiency of smart urban regeneration and smooth the introduction and application of intelligent technology infrastructure. It is difficult to generalize the highly efficient urban spatial structure because it is significantly different depending on the conditions of each city, but it can be approached as a compact city as an example. Danzig and Saaty [1973] explained the characteristics of compact cities by dividing them into the following three: (1) high-rise and high-density dwellings and an urban form with low dependence on automobiles, (2) complex land use and diverse spatial characteristics and (3) social functions for self-sufficiency in everyday life. The OECD [2012] described dense and proximate development patterns, urban areas linked by public transport systems, and accessibility to local services and jobs as characteristics of compact cities. As such, the spatial structure of a compact city is explained by the relationship between density and proximity. In addition, a polycentric compact city planned for complex land use, which is an ideal combination of these characteristics, can be proposed as one of the highly efficient spatial structures for smart urban regeneration. This can bring various benefits from the perspective of smart urban regeneration. Managing the proper density and proximity provides an opportunity to benefit everyone without wasting resources, by inducing the efficient installation of various smart technology infrastructures required during urban regeneration. For example, in the case of an indiscriminately dispersed city, there may be difficulties in the process of constructing a smart technology infrastructure because not only is the infrastructure randomly distributed but so is the population. If the infrastructure is installed in areas with low demand, resource waste and energy consumption for technology operation and maintenance can increase. Conversely, if the infrastructure is not installed in areas with low demand, the problem of service inequality can occur. The intelligent technology infrastructure includes ICT infrastructure and facilities utilizing smart solutions. The introduction and application of intelligent technology infrastructure enables efficient urban management and service provision by collecting and analyzing various data within the city to construct information and utilize it.

'Smart participation' is based on the establishment of smart governance, a cooperative network between various stakeholders engaging in decision-making during smart urban regeneration [Albino *et al.*, 2015]. This enables the smooth progress of smart urban regeneration and the natural acceptance of the bottom-up method through participation and feedback. Therefore, smart participation strengthens cooperation between citizens and the city government, and facilitates the user-oriented planning pursued by smart cities. In particular, ICT is useful in increasing public participation and plays an important role in enabling ICT to have a significant impact on many issues in cities [Åström *et al.*, 2012; Viale Pereira *et al.*, 2017]. In order to maintain smart participation in the process of promoting smart urban regeneration, necessary data must be shared so that anyone can access it. In particular, citizens tend to participate more actively when they have access to useful and complete information from the government; so, equality in access to information is important to achieve smart participation [Bonsón *et al.*, 2015; Mellouli *et al.*, 2014; Viale Pereira *et al.*, 2017].

Finally, there is a need for a 'smart space' that enables sustainable innovation. A shared space should be provided where members can freely participate in education, culture, discussion, information-sharing, experimentation, and finding solutions for a better city. The smart space can be an online platform or a real shared space. Online platforms enable participating entities to perform continuous interaction and feedback without time and place constraints, and such information can be collected in real time. In addition, the shared space within the community can be used as a meeting place between community members and other subjects, a living lab for smart solution development, or a space for start-ups. In a smart space, various members can gather to prioritize urban problems facing their cities and propose appropriate regeneration strategies and smart solutions. For smart urban regeneration, it is necessary for members to accurately recognize the problems faced by each city, identify improvements and directions, and freely collect ideas and collaborate to resolve problems.

14.3 Applications for Smart Urban Regeneration

The applications for smart urban regeneration target projects that realize urban regeneration by incorporating elements of smart cities to solve various problems faced by each city. There are many good examples of smart urban regeneration, but in this chapter, we addressed seven cases that we visited in person. These cases satisfy the three viewpoints of smart urban regeneration summarized in Section 14.2.2, but they were classified by focusing on more meaningful viewpoints. The study consisted of field trips, in-person interviews, and some literature research (brochures, official sites, etc.). As examples of smart environments, we investigated the cases of Munich (Germany), Minden (Germany), and

Seoul (South Korea), which alleviated urban problems or revitalized regions through physical improvement by installing urban facilities. As examples of smart participation, we investigated the cases of the Living Lab in Amsterdam (the Netherlands), where various members voluntarily participated to solve urban problems. As examples of smart spaces, we investigated a makerspace in Munich (Germany) and a smart city innovation center in Seoul (South Korea), which aims to achieve sustainable innovation through shared spaces.

14.3.1 *Smart environment*

14.3.1.1 *Digital information pillar and smart lamppost (Munich, Germany)*

The Neuaubing-Westkreuz/Freiham district in western Munich, Germany is involved in the European Union's (EU) Smarter Together project. Smarter Together is a joint project aimed at improving the quality of life for citizens in a changing city. It focuses on finding the proper balance between ICT technology, citizen participation, and institutional governance to provide a smart and comprehensive solution [European Union, n.d.]. Cities participating in the project are classified into lighthouse cities, follower cities, and observer cities, according to the goals and objectives. Munich is a lighthouse city that serves to provide advanced solutions at the city level, monitor results and provide lessons for the follower cities to use as benchmarks.

Digital information pillars and smart lampposts are installed around Westkreuz station in Figure 14.2. The digital information pillar is a touchscreen kiosk that provides a variety of city information along with all the mobility options available in connection with smart city applications. In addition, the smart lampposts, which are installed at 10-meter intervals to the west of Westkreuz station, consist of adaptive demand control lighting to reduce energy consumption, and M-WLAN hotspots and an integrated sensor that can collect traffic and environmental (contaminants, local climate, etc.) data. Based on this, real-time data of the city are collected and stored in a central repository. The ultimate goal is to provide a service that delivers local information to residents

Figure 14.2. Digital information pillar (a), smart lampposts (b, c, and d), and street view with smart lampposts (e).

through an app and to utilize it for urban development and environmental protection projects.

14.3.1.2 *Beacon Mile (Minden, Germany)*

In the city of Minden, located in North Rhine-Westphalia, 'Digital Minden' is implemented by the Minden Marketing GmbH (MMG). The goal of the project is to establish a network between the actual urban space and the virtual space using various digital technologies such as QR codes, digital marketplaces, and the Minden app augmented reality [Minden Marketing GmbH, n.d.]. In 2014, a free internet network was installed in the center of Minden; and in 2016, the Beacon Mile was introduced for the first time in Germany. The beacon, defined as a light or signal indicating a location, is a short-range communication device that transmits appropriate data or information to users within a range of 50 meters to 70 meters. Figure 14.3 shows a shopping street in Minden city.

Minden city sought to revitalize the region by actively utilizing existing local resources such as historic buildings and shopping centers considering regional conditions and devised a plan to introduce the appropriate beacon mile. Minden city has installed 75 beacon mile transmitters in the streets from the cathedral in the old town, the old town hall, and the shipyard along the Weser River in Figure 14.4. This is a street that attracts many tourists due to the preservation of

Figure 14.3. Shopping street view in Minden city.

Figure 14.4. Wi-Fi AP and beacon installed on the street.

Renaissance-style buildings including cathedrals. The Beacon Mile provides tourists with guides for Minden tourism and visitors with benefits related to shop use, and serves as a marketing platform for retail and restaurant owners. The transmitter sends a push message with information related to a facility in an app installed on a smartphone within a radius of 70 meters. Walking through the "beacon mile section", where transmitters are installed, information related to museums, restaurants, and bus stops is received on a smartphone in real time. In addition, when selecting a menu such as beer or coffee through the app, it guides a user to the nearest restaurant or cafe from the current location and provides various services such as restaurant information and discount coupons.

14.3.1.3 *Smart mobility green station and smart moss tower (Seoul, South Korea)*

The Smart Mobility Green Station and Moss Tower project was promoted in the Magok district of Seoul, South Korea in Figures 14.5 and 14.6. These projects are part of a smart city service project for residents that are being promoted by the Smart City Business Group of Seoul

Figure 14.5. Smart mobility green station.

Figure 14.6. Smart moss tower and QR code for AR service.

Housing and Communities Corporation of the Seoul Metropolitan (hereinafter referred to as "SH").

Magok District is a section in which the concentration of ultrafine dust, or fine particulate matter (PM 2.5) is particularly high due to heavy traffic. Therefore, SH developed the Smart Mobility Green Station and Smart Moss Tower to improve this environment. The smart mobility green station is the result of harmonizing Mobility as a Service (MaaS) and technology to reduce fine dust and heat waves. This seeks to solve the problems of mobility and fine dust in urban living areas at the same time. The station is designed to be able to allow scooters to be returned, and charge an electric scooter charging and locking system simultaneously, and it is the first system to provide a smart mobility service in Korea. This will provide first- and last-mile services to employees and residents of approximately 190 companies at five subway stations in Magok District, and it can be expected to reduce the emission of fine dust by restraining the use of unnecessary vehicles. An additional function of the station is to install sensors to monitor fine dust, temperature, humidity, and wind. Sensed data are transmitted to the server and used to provide real-time information to citizens. In addition, green roofs and vertical greening are provided to reduce the temperatures near the station in the summer and purify the air environment through the adsorption of fine dust.

The Smart Moss Tower is a smart solution developed to reduce the fine dust in cities. It can provide a cloud-based service that combines IoT technology with the characteristics of moss, one of the plants that can easily absorb the pollutants generated in an urban environment. The tower is circular with a height of approximately two meters, and uses moss plants and natural and artificial winds to maximize the absorption effect of fine dust within a radius of 50 meters and purify the air. In addition, the irrigation system is designed to automatically supply the water required for the moss and to operate the system in connection with real-time climate information. Real-time information on fine dust reduction is managed on a cloud basis, and it is developed to enable remote automatic control with the smart city operating center and smartphone app. It also provides a simple augmented reality (AR) experience service that allows people to check the status of the concentration of fine dust around the area in real time.

14.3.2 *Smart participation*

14.3.2.1 *Marineterrein Amsterdam (Amsterdam, Netherlands)*

In 2009, the Amsterdam City Hall, businesses, and academia created the Amsterdam Smart City (ASC) program to implement a smart city. Currently, experts and citizens are promoting various projects on digital cities, citizens and living, energy, circular cities, mobility, and smart city academics. The ASC uses the Living Lab method that finds solutions through experiments by applying appropriate technologies to existing urban problems. In the processes of Living Lab projects, the city of Amsterdam provides an online–offline platform that connects startups to develop technology, experts for advice, and civic groups; and residents select the appropriate technologies and how to use them.

Currently, more than 40 Living Lab facilities have been established in Amsterdam, and one of the representative living labs is the Marineterrein Amsterdam. In 2011, the Ministry of Defense decided to relocate the Marines from the Marineterrein area, and the central and local governments began planning the Marineterrein Amsterdam in 2013. Development began in 2015 when the site was changed to being owned by the central government. The building that used to be a naval base was regenerated as a Living Lab facility, where startups and experts can work to solve problems such as health, housing, transportation, and energy problems facing the city. In the facility, individuals or organizations with similar interests and related knowledge about problems recognized by the state, issues raised by citizens, and ideas proposed by companies can gather and actively exchange opinions; and they can also receive specialized knowledge and networks. The project also has a "plug-and-play" infrastructure for data, electricity, water and sewage, both underground and above ground, providing a perfect place for testing and experiments.

Representative Living Labs conducted at the Marineterrein Amsterdam include the 'Boardwalk', developed by a startup and the 'smart roof 2.0', which started with a proposal from a nearby hotel restaurant in Figure 14.7. The 'Boardwalk' is a bridge that can cross the area across the inner port adjacent to the Marineterrein Amsterdam, allows free design by assembling an approximately 10-meter module and prevents contaminants from passing into the inner port. The 'smart roof 2.0' is a crop cultivation project using the roof of a building. When the system detects that the plant needs water via 50 automatic sensors installed on the roof floor, it gives pre-stored rainwater to the plant and lowers the temperature of the building at the same time. Additionally, the Living Labs are conducting life-friendly experiments such as electric bicycles linked to transportation cards, restaurants that do not create food waste, and greenhouses that use animal waste as an energy source.

Figure 14.7. "Boardwalk" built in the inner port of Amsterdam.

14.3.2.2 *De Ceuvel (Amsterdam, Netherlands)*

De Ceuvel is a Living Lab facility that reconstructed the site of an abandoned shipyard in north Amsterdam, the Netherlands in Figure 14.8. It has been providing workspaces, performance halls, and cafes by renovating 17 houseboats since 2014. Due to the low rent, De Ceuvel brings together city experts from various fields, including founders and eco-friendly consulting firms, to revitalize the region.

De Ceuvel's tenant companies and organizations are conducting experiments to ultimately transform into a sustainable smart city through urban regeneration experiments, such as self-sufficient living communities, energy independence, self-procurement of food materials, and household waste recycling (food waste, etc.). Representatively, there are contaminated soil remediation and compost toilet projects. Contaminated soil remediation is a project in which plants are planted to purify contaminated soil in the De Ceuvel area, which has been used as a shipyard in the past, in an eco-friendly way. The compost toilet project is an experiment that turns the manure collected from the De Ceuvel toilets into compost for growing crops. All 17 boat offices have water-free compost toilets that dry compost to extract phosphate from urine and make it into fertilizer. After filtering pathogens or metal components from the compost, they are sent to the greenhouse to be used

Figure 14.8. Panoramic view of Amsterdam De Ceuvel.

for vegetable cultivation. Recently, the project has promoted Blockchain-based energy demonstration by installing solar power in buildings and an experiment that uses the generated energy as a local currency (reward).

14.3.3 *Smart space*

14.3.3.1 *Makerspace (Munich, Germany)*

The 1,500 square meters makerspace in Figure 14.9, located at the Technical University of Munich (TUM), was started in 2015 by an

Figure 14.9. Maker space at the Technical University of Munich.

investment of 20 million Euros by BMW, an automobile manufacturer headquartered in Munich. TUM operates this space as a high-tech workshop equipped with 3D printers, laser cutting machines, and ultra-high-pressure water flow cutting machines. The makerspace provides an environment for making prototypes using digital technology. There are more than 30 resident employees, and they play various roles such as educating users on how to operate the machines. This space can be used by anyone, including students at the TUM, prospective founders, craftsmen, and corporate executives such as Bosch and Siemens (the number of users is more than 1,000 per year). In addition, this space has become a platform for composing project teams, allowing teams to be freely formed and disbanded according to projects; and a variety of communities can be formed. Smart spaces, such as makerspaces, share the machines, facilities, and spaces necessary to realize ideas that can sustain smart cities or develop real products. This will allow any member of society to freely contribute to maintaining a smart city and ultimately provide a basis for developing the region more 'smartly'.

14.3.3.2 *SH Smart city innovation center (Seoul, South Korea)*

The 'SH Smart City Innovation Center' in Figure 14.10 located in Seoul, South Korea is a space prepared by the SH to revitalize the smart city business. It was established for the purpose of establishing a smart business network among various partners, such as citizens and innovative companies, and developing innovative solutions and services that can be practically applied to their business districts. This center operates an innovation business center for monitoring the smart city services applied to the business district of SH, conducting integrated data management, and housing multipurpose meetings between partners. It also provides an innovation solution lab for living labs, smart content production studios, and education. In particular, the operation of the 'innovation solution lab' supports various partners such as citizens and innovative companies to develop smart city services and solutions, create and distribute creative smart content, and support offline

Figure 14.10. Interior and exterior view of the SH smart city innovation center.

spaces. The four solution labs currently in operation are as follows: (1) the 'smart AR street lab' for creating and sharing smart augmented reality content, (2) the 'smart green street lab' to reduce fine dust and create eco-friendly smart streets, (3) the 'smart energy IT lab' to build a smart energy platform suitable for SH business and (4) the 'smart green mobility lab' to activate eco-friendly MaaS. Through this, the participants have achieved tangible results such as the "smart mobility green station" and "smart moss tower" mentioned in the smart environment part. It is meaningful that public institutions have independently created a space for collaboration between working-level staff, citizens, innovative companies, and external experts in order to understand the needs of citizens living in the business district and develop smart services appropriate to them.

14.4 Conclusion

'Smart city' is currently the most popular key term, but it is very difficult to form a consensus on its definition. In this chapter, a smart city was regarded as a combination of urban planning that connects technological progress and people–information–knowledge–city. In addition, a smart city was thought to be a city that aims to operate the city efficiently and provide high-quality services. When introducing smart city planning as a new paradigm of urban regeneration, the three dimensions of the smart environment, smart participation, and smart space are essential for sustainability. Moreover, elements such as monitoring and feedback, solution development and application, and policy and system should be supported.

In this chapter, it was confirmed how the essential dimensions and supporting elements of smart urban regeneration are reflected based on application cases (see Table 14.1). In the case of the smart environment, a service that can increase the convenience of living was mainly provided by reflecting the requirements of the existing city and its citizens. In the process, appropriate solutions considering regional conditions were planned, or plans to fully utilize local resources were proposed to contribute to regional vitalization. Smart participation provided opportunities for individuals or organizations to solve urban problems and maintain sustainability, and enabled the exchange of opinions, the sharing of expertise, and the formation of networks. In addition, abandoned buildings or sites were regenerated to revitalize deteriorated areas. Smart spaces are aimed at the sustainability of smart city services by providing spaces where various participants can communicate and collaborate to achieve smart innovation.

As the 4th Industrial Revolution approached, the development of technology and the resulting industries changed rapidly. In the future, technology will develop faster than before; and with each development, the previous technology will be relatively out-of-date. Therefore, we need to raise the question of whether integrating the latest technologies into cities can guarantee the sustainability of smart cities. From the case study introduced in this chapter, we can confirm that the integration of the latest technologies is not the only way to lead a smart city. The

Table 14.1. Application cases.

Category	**Cases**	**Roles**	**Benefits**
Smart environment	Munich (Germany): Digital information pillars and smart lampposts	– Providing various city information – Providing M-WLAN hotspots – Collecting traffic and environmental data	– Providing services that can enhance the convenience of living by reflecting the needs of citizens – Utilizing existing local resources considering regional conditions – Regional revitalization through suitable solution and planning
	Minden (Germany): Beacon Mile and Wi-Fi devices	– Providing free Wi-Fi – Providing local information and guides – Providing services linked to retail stores	
	Seoul (South Korea): Smart mobility green station and smart moss tower	– Electric scooter rental – Collecting environmental data and providing real-time information to citizens – Reducing the emission of fine dust and air purification – Facility management through automatic remote control	
Smart participation	Marineterrein Amsterdam (Netherlands): Boardwalk, smart roof 2.0, etc.	– Participation of individuals and organizations interested in urban issues and ideas	– Providing participatory opportunities for individuals or organizations with the same goal

(*Continued*)

Table 14.1. (*Continued*)

Category	Cases	Roles	Benefits
	De Ceuvel (Netherlands): Contaminated soil remediation and compost toilet projects, etc.	– Participation of experts, companies, and organizations in various fields to form a sustainable city	– Exchange of opinions among participants, sharing specialized knowledge and networking – Operating a Living Lab aiming at solving urban problems – Regeneration of abandoned buildings or sites to revitalize the deteriorated area
Smart space	Munich (Germany): Makerspace	– Providing high-tech machinery – Staff for training and support – Providing a platform for the community	– Creating a space for collaboration between staff, citizens, innovative companies, and experts
	Seoul (South Korea): Smart city innovation center	– Monitoring of smart city services – Integrated data management – Providing spaces for collaboration, creative activity, and education	– Providing spaces that support machines, facilities, and education to support the sustainability of smart city services

examples of smart urban regeneration introduced in this chapter do not highlight the state-of-the-art technologies commonly mentioned in smart city planning. They only adopt smart technologies that are effective in making decisions in the process of solving urban problems and suitable for urban planning. These ideas are more useful in existing cities than in planning a new city. In the case of new urban development, a stable smart city can be created by planning and designing a smart city from the initial stage of development. However, existing cities must further consider the urban infrastructure developed and the needs of citizens.

The essence of smart urban regeneration is to understand the conditions and characteristics of each city and to provide high-quality services to users through the application of efficient smart technologies and solutions accordingly and to provide integrated management to managers. Therefore, for smart urban regeneration, it is necessary to provide shared spaces that can support the management of the appropriate spatial structure and infrastructure for each city and the participation of various members. Based on this, smart technologies, solutions, and plans that reflect the spatial structure and infrastructure conditions and the needs of citizens living in the city should be established.

References

Albino, V., Berardi, U., and Dangelico, R.M. (2015). Smart cities: Definitions, dimensions, performance, and initiatives, *Journal of Urban Technology*, *22*(1), 3–21. URL: https://doi.org/10.1080/10630732.2014.942092

Angelidou, M. (2015). Smart cities: A conjuncture of four forces, *Cities*, *47*, 95–106. URL: https://doi.org/10.1016/j.cities.2015.05.004.

Anthopoulos, L., and Fitsilis, P. (2010). From digital to ubiquitous cities: Defining a common architecture for urban development, *Proceedings—2010 6th International Conference on Intelligent Environments, IE 2010*, 301–306. URL: https://doi.org/10.1109/IE.2010.61.

Åström, J., Karlsson, M., Linde, J., and Pirannejad, A. (2012). Understanding the rise of e-participation in non-democracies: Domestic and international factors, *Government Information Quarterly*, *29*(2), 142–150. URL: https://doi.org/10.1016/j.giq.2011.09.008.

Bakıcı, T., Almirall, E., and Wareham, J. (2013). A Smart City Initiative: The Case of Barcelona, *Journal of the Knowledge Economy*, *4*(2), 135–148. URL: https://doi.org/10.1007/s13132-012-0084-9.

Bibri, S.E. (2019). On the sustainability of smart and smarter cities in the era of big data: An interdisciplinary and transdisciplinary literature review,

Journal of Big Data, *6*(1), 1–64. URL: https://doi.org/10.1186/s40537-019-0182-7.

Bolici, R., and Mora, L. (2015). La rigenerazione urbana nell'era digitale: Come sviluppare strategie Smart City in città europee di grandi dimensioni. Urban reneneration in the digital era: Houto develop Smart City strategies in large european cities (Urban regeneration in the digital era: How to develop Smart City strategies in large European cities), *Techne*, *10*, 110–119. URL: https://doi.org/10.13128/Techne-17507.

Bonsón, E., Royo, S., and Ratkai, M. (2015). Citizens' engagement on local governments' facebook sites. An empirical analysis: The impact of different media and content types in Western Europe, *Government Information Quarterly*, *32*(1), 52–62. URL: https://doi.org/10.1016/j.giq.2014.11.001.

Chourabi, H., Nam, T., Walker, S., Gil-Garcia, J.R., Mellouli, S., Nahon, K., … Scholl, H.J. (2012). Understanding smart cities: An integrative framework, *Proceedings of the Annual Hawaii International Conference on System Sciences*, 2,289–2,297. URL: https://doi.org/10.1109/HICSS.2012.615.

Couch, C., Sykes, O., and Börstinghaus, W. (2011). Thirty years of urban regeneration in Britain, Germany and France: The importance of context and path dependency, *Progress in Planning*, *75*(1), 1–52. URL: https://doi.org/10.1016/j.progress.2010.12.001.

Dantzig, G.B., and Saaty, T.L. (1973). *Compact City: A Plan for a Liveable Urban Environment*. WH Freeman.

DESA, U. (2019). *World Urbanization Prospects 2018: Highlights (ST/ESA/SER. A/421)*.

European Union. (n.d.). *Smarter Together*. URL: https://www.smarter-together.eu/about/objectives. Accessed: November 12, 2020.

Mellouli, S., Luna-Reyes, L.F., and Zhang, J. (2014). Smart government, citizen participation and open data, *Information Polity*, *19*(1–2), 1–4. URL: https://doi.org/10.3233/IP-140334.

Minden Marketing GmbH. (n.d.). *Digital Minden*. URL: https://www.minden-erleben.de/tourismus/index.php/en/digital-minden/approaches-and-goals. Accessed: November 23, 2020.

Nam, T., and Pardo, T.A. (2011). Conceptualizing smart city with dimensions of technology, people, and institutions, *ACM International Conference Proceeding Series*, (June), 282–291. URL: https://doi.org/10.1145/2037556.2037602.

Organisation for Economic Co-operation and Development. (2012). *Compact City Policies: A Comparative Assessment*. OECD.

Sharifi, A. (2019). A critical review of selected smart city assessment tools and indicator sets, *Journal of Cleaner Production*, *233*(1), 1,269–1,283. URL: https://doi.org/10.1016/j.jclepro.2019.06.172.

Sharifi, A. (2020). A typology of smart city assessment tools and indicator sets, *Sustainable Cities and Society*, *53*, 101936. URL: https://doi.org/10.1016/j.scs.2019.101936.

Shen, L., Peng, Y., Zhang, X., and Wu, Y. (2012). An alternative model for evaluating sustainable urbanization, *Cities*, *29*(1), 32–39. URL: https://doi.org/10.1016/j.cities.2011.06.008.

Viale Pereira, G., Cunha, M.A., Lampoltshammer, T.J., Parycek, P., and Testa, M.G. (2017). Increasing collaboration and participation in smart city governance: A cross-case analysis of smart city initiatives, *Information Technology for Development*, *23*(3), 526–553. URL: https://doi.org/10.1080/02681102.2017.1353946.

Yigitcanlar, T., O'Connor, K., and Westerman, C. (2008). The making of knowledge cities: Melbourne's knowledge-based urban development experience, *Cities*, *25*(2), 63–72. URL: https://doi.org/10.1016/j.cities.2008.01.001.

Yigitcanlar, T. (2009). Planning for smart urban ecosystems: Information technology applications for capacity building in environmental decision making, *Theoretical and Empirical Researches in Urban Management*, *4*(3), 5–21.

© 2023 World Scientific Publishing Company
https://doi.org/10.1142/9789811257186_0015

CHAPTER 15

The Smart City in Developing Countries: Experiences from Brazil

Daniel Esashika,[*] Gilmar Masiero,[*] Yohann Mauger[†,‡]

[*]*Department of Management, Faculty of Economics, Administration and Accounting, University of São Paulo, São Paulo-SP, São Paulo, Brazil*
[†]*Department of Management, Haile/US Bank College of Business, Northern Kentucky University, Highland Heights, Kentucky, USA*
[‡]*Corresponding author, Email: maugery1@nku.edu*

15.1 Introduction

Cities have become the major centers of human life. Since 2007, urban population has surpassed rural population, and it is estimated that, in 2050, 70% of the world population will live in cities [UN, 2018]. We should note that developing countries have triggered a large part of this movement. According to Cohen [2006], there are at least 400 cities globally with over one million inhabitants, and 70% of these are in less developed countries. Considering developing countries, Latin America stands out for having the fastest urbanization process in the last four decades [Fernandes, 2007].

Meanwhile, this accelerated urbanization process has brought new challenges. Despite the favorable impact of urban life compared to rural life, a greater access to goods and services remains limited by the cities' ability to support population growth [Shen *et al.*, 2012]. The disordered urban expansion drastically affects rural areas and ecological habitats, impacting natural processes [Zeng *et al.*, 2016]. This is also reflected in health risks arising from the harmful effects of air pollution in the environment [Mannucci and Franchini, 2017], poor basic sanitation [Zhou *et al.*, 2018], as well as traffic and mobility problems [Pojani and Stead, 2015].

Another aspect highlighted by some researchers is the need to reduce socioeconomic inequalities resulting from rural migration, which raises the discussion about broader aspects of sustainability [Dempsey *et al.*, 2011]. There is a tendency for the service and industrial sectors to expand in these cities, directly affecting the demand for education [Srivastava *et al.*, 2015] and energy, preferably from renewable and clean sources [Hu *et al.*, 2018].

To face these challenges, the United Nations (UN) published the New Urban Agenda after the Quito conference [UN, 2016]. In this context, the agenda ratifies the global commitment to promoting sustainable urban development and contributes to implementing the 2030 Agenda for Sustainable Development. In particular, it supports the Sustainable Development Goal 11, which sets the goal for cities and human settlements to become inclusive, safe, resilient, and sustainable [UN, 2015].

In this context, smart cities are seen as a promise of contemporary urbanism, in which the use of new technologies would boost the capacities of local inhabitants and their governments to improve the living conditions in the cities [Gil-Garcia *et al.*, 2015]. Although initially, there is a focus on technologies such as the Internet of Things (IoT) devices and Artificial Intelligence (AI), in the last decade, smart cities have moved towards the development of human potential in their urban life [Kummitha and Crutzen, 2017].

Given this panorama, this chapter explores smart cities' experiences in Brazil, a country with two megacities—São Paulo and Rio de Janeiro—and 17 municipalities with over one million inhabitants [IBGE, 2020]. In Brazil, about 83% of the population lives in cities [Fernandes, 2007]. However, 26 million people living in urban areas do not have access to water; 83 million are not connected to sewage systems; and 70% do not treat the collected sewage [IBGE, 2020].

Other figures suggest that, despite the often-long distances involved, 52 million Brazilians walk to work, due to high public transportation costs. In addition, regarding housing, a national housing deficit of 7.9

million units is estimated, while in the meantime, 5.5 million units are vacant properties. Finally, Brazil faces the challenge of the "inner burbs", or suburban settlements formed in the first decades of industrial growth (1950–1980) characterized by informal mechanisms of access to land, and without basic urban structures [Ward *et al.*, 2015].

These numbers show some of the big challenges that smart cities tackle. Most of these challenges are categorized into four items in this chapter. The first section contextualizes aspects of urbanization in developing countries and gives a brief introduction to the Brazilian scenario. The second section presents a discussion of recent urban policies and recent initiatives in Brazil to develop smart cities. The third section deals with Brazilian smart cities' experiences to fulfill the principles set out in the New Urban Agenda and the UN 2030 Agenda. Finally, the fourth section includes considerations and recommendations about the suitable development of smart cities in Brazil. The chapter awakes some insights to reflect on smart city policies geared towards developing countries.

15.2 From City to Smart Cities in Brazil

Following different "models" of economic development or industrialization, countries worldwide face many troubles in their urbanization experiences. Since the 1950s, the Western world and, more recently, the Eastern part of the world have been rapidly spreading their cities' infrastructure. At the beginning of this century, more than 50% of the world's population starts living in cities for the first time. Massive migrations from rural to urban areas are challenging government authorities in the developing countries. While some countries are urbanizing the rural areas, others are receiving many migrants and expanding their large metropolitan regions' infrastructure.

Latin American countries with no exceptions opted, more than in most of the other regions, to expand the urbanization of their historical cities located in their coastal region. According to a Bank report [BBV, 2017], after a sharp growth in the 1950s, Latin American urbanization has slowed down its pace. In the coming decades, the urbanization is expected to grow at a slower pace than the world average. This same report mentions that despite the regions showing high urbanization rates, Latin American countries struggle with low income, capital, and productivity. For the banks, while urbanization shows some positive associations with digitalization, most cities in the regions have low physical, social, and digital connectivity, which leads to a high fragmentation.

The Brazilian case, recent Brazilian urban policy, is marked by coordination difficulties related to the federalist system. According to the Brazilian Federal Constitution of 1988, there is a sharing of responsibilities among the federal entities regarding urban law. The constitution establishes greater autonomy for municipalities and makes the master plan mandatory for cities with more than 20,000 inhabitants [Macedo, 2008]. This was undoubtedly a milestone for Brazilian cities' planning and governance, in recovery from the disorderly growth and precarious urban services characteristic in the last 50 years of the 20th century [Rolnik, 2013].

Lefebvre [1967] and Harvey [1973] advocated new legal diplomas and urban laws such as Law No. 10,257/2001 in Brazil—recognized as a City Statute. The City Statute is seen as a complement to the Urban Policy chapter laid down in the Federal Constitution. Among the statute's major innovations, instruments are created to rationalize the land and democratize urban management.

Santos Junior and Montadon [2011] pointed out the low effectiveness of the statutes' instruments, which are foreseen in the master plans but are not actually implemented. This result is attributed to the lack of the plan's details, the lack of technical and institutional capacity of the municipal government departments, as well as the difficulty of articulating urban policies in regions bordering metropolitan regions. Taking these issues into account, the Brazilian National Congress enacted the Metropolis Statute, the Law No. 13,089/2015.

Among the major advances of this new law is the conceptualization of metropolitan areas and metropolises, previously defined according to political and unclear criteria. The Metropolis Statute reinforced governance mechanisms, the public services to be served by metropolitan management, and the means for the social control of these activities [Araujo *et al.*, 2016]. In this sense, the Metropolis Statute strengthens society's participation and control in decisions about urban management.

Therefore, we can say that Brazilian cities' planning and governance instruments have strengthened in the last three decades, supported by successive advances in legislation.

Concerning the technological development and expansion of the country's connectivity, under the Growth Acceleration Program (Programa de Aceleração do Crescimento, or PAC), the Digital Cities (Cidades Digitais) started to develop several large infrastructure projects in 2012. At the time, a significant part of the Brazilian population did not have access to the Internet. To solve the problem, the government visualized the possibility of building an Open Access Metropolitan Area Network (MAN), like what was accomplished under the eEurope initiative [Rezende *et al.*, 2014]. However, this initiative was not accomplished. The program presented a series of weaknesses, such as the insufficiency in the design of public policy aiming at fundamental pillars for digital inclusion (infrastructure, literacy, and content) and the lack of coordination among the different sectors.

In 2019, the Brazilian Program for Sustainable Smart Cities replaced the Digital Cities program. The new program aims to establish guidelines, indicators, and thematic axes for evaluating ongoing smart city initiatives. It is inspired by the documents "Smart Sustainable Cities Maturity Model" and "Assessment Framework for Digital Transformation of Sectors in Smart Cities" promoted by the International Telecommunication Union (ITU). To mitigate the problems identified in the Digital Cities program, a Chamber of Smart Cities was created to promote discussions involving governments, academia, industry, the private sector, and civil society representatives.

Concomitantly, two other policies consolidate this new program: (1) the National IoT Plan and (2) the Brazilian Charter for Smart Cities. Published in 2019 and based on the study "Internet of Things: An action plan for Brazil" (Internet das Coisas: um Plano de Ação para o Brasil) [BNDES, 2018]. The National IoT Plan established an action directed to four strategic sectors—cities, health, agribusiness, and industry. More particularly, concerning cities, the plan foresees the adoption and development of technologies for urban mobility, public safety, energy efficiency and sanitation, and innovation by adopting locally developed solutions. While the National IoT Plan focuses on technological and economic aspects, the Brazilian Charter for Smart Cities articulates an urban development agenda that includes more aspects. The document, which is under virtual public consultation since 2020, was prepared by a team of more than 200 specialists under the Ministry of Regional Development's coordination, the Ministry of Science, Technology, and Innovations, and the Ministry of Communications. The charter presents the definition of a smart city in the Brazilian context, the public agenda for the development of smart cities, an index of ongoing initiatives, and support for local actions.

The charter establishes eight strategic objectives for smart cities in Brazil: (1) integrating digital transformation and sustainable urban development, (2) providing quality internet access, (3) offering security, transparency, and privacy, (4) establishing innovative and inclusive models of urban governance, (5) promoting local economic development in digital transformation, (6) stimulating financing and instruments, (7) fostering a massive and innovative education movement and (8) developing models to assess the impacts of digital transformation on cities. Its orientation towards smart cities goes far beyond the city's technological vision, emphasizing aspects related to human development, digital citizenship, and participatory governance.

Within the legislative power, an extensive study was carried out by the Chamber of Deputies and published as a book titled *Smart Cities: a human and sustainable approach*, or *Cidades Inteligentes: Uma Abordagem Humana e Sustentável* [CEDES, 2020]. The book summarizes the views of Brazilian experts and representatives of all entities of the federation and municipal secretaries responsible for the implantation of smart cities. The book is a landmark on the development of Brazilian smart cities and guides the main legal changes necessary to enable cities' digital transformation.

Brazil has some international commitments related to the urban issue—the 2030 Agenda and the New Urban Agenda. These commitments foster a global orientation towards local communities' sustainability. In particular, it focuses on issues related to poverty, social inequality, environmental sustainability, as well as social engagement and participation [Carpenter, 2015]. Brazil's commitment to implementing international agendas is strong due to the past Brazilian leading role at the Habitat Conferences in Vancouver (1976), Istanbul (1996), and Quito (2016). In this sense, the legal frameworks on urbanization in Brazil incorporated the values and actions proposed by the UN agendas.

15.3 Smart Initiatives in Brazilian Cities

Promoting smart cities at the national level and at the Brazilian federal government level started just a few years ago. For more than a decade, several cities have taken individual initiatives to develop their own plans. This contrasts with the situation in North America, Europe, and Asia. In these countries, the development of smart cities started in the first decade of the 21st century, with observable results today. In this section, we highlight some Brazilian cities which implemented intelligent experiences.

Looking at the regional capitals and the Federal District, we observe some plans for developing a smart city. In this sense, we can say that the concept is widespread among the main public managers in the country. However, sometimes there is a more technological orientation, formalized through master plans for smart city technologies, while a more holistic approach covering human, cultural, and social aspects is lacking.

Although there are no specific plans for smart urban development, isolated initiatives highlighting aspects associated with smart cities prevail, such as electric vehicles and intelligent public lighting. The following items present some of the Brazilian experiences aligned with the vision for cities in the New Urban Agenda. This agenda challenges cities and human settlements to be fair, safe, healthy, physically, and economically accessible, but also resilient and sustainable to foster prosperity and quality of life for everyone [UN, 2016].

15.3.1 *Social justice*

Education is perhaps the factor that hinders the smart cities' development in Brazil the most, as significant inequality characterizes the educational system. The latest data from the Programme for International Student Assessment (PISA) 2018 show that part of the Brazilian educational system is compatible with Organisation for Economic Co-operation and Development (OECD) countries, emphasizing federal public schools and private schools in urban environments. However, at the same time, the country performs below the OECD average in the sample of state and municipal schools, especially those in rural environments. In other words, there is a part of Brazil that receives high level education, comparable to the best global examples of Estonia and Finland, while another part receives some of the lowest education. Undoubtedly, dealing with inequality in the educational system and its consequences will be one of the key challenges for the Brazilian cities to overcome.

In addition, while investigating the digitalization, via the penetration of the Information and Communication Technology (ICT) in the Brazilian society, Nishima *et al.* [2017] identified that the digital exclusion had decreased dramatically between 2005 and 2013. They pointed out spaces for policies of wide access to ICT devices and broadband internet access. Besides, they attributed digital illiteracy as the principal cause of digital exclusion in the country, especially among the elderly. It should be noted that these criticized aspects were widely discussed during the quarantine imposed by the COVID-19 pandemic, where a significant portion of the community was without access to basic services (for example, education) due to poor access to the Internet or a lack of ICT devices at home.

The wealthiest city of the country, the capital of São Paulo (SP), has a complete program of digitalization. In the WiFi Livre SP program, via a partnership between the city and Google, there is a connectivity network with more than 620 access points spread across the city. Thinking beyond the provision of infrastructure, the city of São Paulo has a network of Digilabs, which are centers focused on teaching digital technologies and introducing new users. There is also a network of Fablabs to stimulate the production side, in which users can make three-dimensional (3D) prints, drones, and carpentry, aimed at developing the ability to build and change different objects, i.e., the so-called maker culture.

The city of Goiânia (GO) has also worked on digital literacy, targeting different audiences. The Makers Project, aimed at public schools, seeks to arouse students' interest in developing robotics and programming solutions. It is an initial stimulus that encourages training in the areas of STEM—Science, Technology, Engineering, and Mathematics. Besides, the program emphasizes teachers' technological training, with technical support from the Federal University of Goiás (UFG). For this audience, the municipal managers of Goiânia articulated partnerships with social services organizations maintained by commercial and industrial associations. These entities have a catalog of technical courses, and the Goiânia city hall, through its employment agency, articulates the training of professionals according to local companies' needs.

Preparing the population to experience a digital society is not restricted to the education sector and the training of professionals. A recent report by the Brazilian Association of Information and Communication Technology Companies (BRASSCOM) points to a deficit in the number of professionals trained annually and the difficulty of filling vacancies currently available. The situation is expected to become more critical in the coming years, considering that the Association projects an expansion of demand for another 420,000 new professionals by 2024. A similar situation occurs in engineering sectors such as oil, gas, and mining, in which there is a great demand for professionals trained in information technologies.

The experience of the Belo Horizonte City Hall (BH) is also interesting because its wide digital literacy program includes older adults. For the entire population to enjoy smart cities' benefits and governments' digital transformation, it is necessary to include digital nonnatives. In this context, the city of Belo Horizonte promotes courses in nursing homes to introduce the elderly to the digital world. The OECD recently highlighted the advancement of the digital transformation in Brazil, ranking the country in 16th place globally. However, this progress is ineffective without citizens, especially the most vulnerable population, taking advantage of new solutions.

15.3.2 *Security*

Brazil is one of the most violent countries, with a homicide rate of 27.1 per 100,000 inhabitants. A large part of the violence and crime rates are related to drug and arms trafficking coordinated by organized groups such as gangs and militias [Miraglia, 2016; Wilding, 2013]. In some cities, there is a conflict of power between the state and these parallel criminal groups. The local population, especially the poor communities, also question and doubt the work of the police and prefers the protection of these criminal groups [Germany, 2014].

The best-known Brazilian experience of controlling violence and crimes is the Rio Operations Center (COR) of the Rio de Janeiro City Hall (RJ). Established in 2010 to organize the Rio 2016 Olympic Games, COR integrates around 30 bodies to monitor the city's operations and minimize unforeseen events on the citizens' routines. Among the emergency occurrences alerted by COR are heavy rains, landslides, and traffic accidents. Besides, a network of over 800 security cameras and sensors scattered throughout the city can offer information on crimes, also alerted by sensors that detect unusual noises and environmental problems, such as the clogging of sewer holes and air quality.

Other cities have implemented similar systems. The Operation and Security Intelligence Center (COI) of Salvador (BA) articulates operations by the police and the fire department. The COI has technological monitoring, communication, and intelligence instruments, such as more than 1,000 security cameras, and a sophisticated facial recognition algorithm. It proved the system's effectiveness during the carnival in Salvador when it recognized a fugitive from the police dressed as a woman in the crowd. The result is impressive, considering that the Salvador Carnival is one of Brazil's busiest festivals, with over three million participants annually. Porto Alegre (RS) implemented an Integrated Command Center (CEIC), with pilot tests in Canoas, part of its metropolitan region. More than 30 sensors were installed to detect noise, such as gunfire in the neighborhoods with high homicide rates. The adoption of this new technology and security actions helped reduce the homicide rate by 38%.

15.3.3 *Health*

The 1988 Brazilian constitution established the universal right to health and assigned the State's responsibility to provide the service. In this context, the unified health system (Sistema Único de Saúde, or SUS) was created in 1990. Over three decades of its existence, the SUS has contributed to improving health services and equity [Souza *et al.*, 2018]. There is a special recognition of its participation in the improvement of life expectancy at birth rates. However, the rates of satisfaction with the health services offered by the SUS have significantly reduced in the last decade [Passero *et al.*, 2016].

Besides, the SUS's sustainability is threatened by the fiscal austerity measures implemented in 2016, the low collaboration between the private and public sectors, persistent regional inequalities, and recent environmental and educational policies [Castro *et al.*, 2019; Massuda *et al.*, 2018]. Therefore, the SUS faces the challenge of addressing a growing and unsatisfied demand for health services, with proportionally fewer resources than it had in previous decades. In this scenario, implementing new work processes and technologies become a critical factor.

A promising path to the health system can be seen in the Family Clinic's experience in the city of Jundiaí (SP). The experience is an example of the technical possibilities associated with efficient process management. The region is recognized for being a regional reference for in-hospital treatments of high complexity. However, it presented poor indicators associated with primary, preventive, and cheaper care. As a way of changing the scenario, they implemented a Family Clinic model, which envisaged the adoption of complete teams of health professionals (doctors, dentists, nurses, nursing technicians, and community health agents) for each group of 5,000 citizens.

Also, there was a screening of medical care demands in routine and specialized consultations, which rationalized human and material resources. The result of this was a 75% reduction in referrals to medical specialties. This change made it possible to accelerate the access to specialized medical care specialist (i.e., dermatologist, gastroenterologist, ophthalmologist, etc.) from 120 days to 48 days. To better target the resources at the city level, a technological health data management system was created, enabling the mapping of each micro-region's demands. Finally, an appointment scheduling app was developed, helping users evaluate their experience with the medical provider as well as tracking problematic health cases directly to the municipal health secretary's office.

Another initiative, also based on the use of apps, is "Saúde Já" from Curitiba. In 2017, the app was recognized by the Latam Smart City Awards 2018, which selects the best smart cities' ideas in Latin America. The app had the two following features when launched: (1) virtual vaccination card and (2) scheduling appointments and exams. Subsequently, other features were added to assist pregnant women in prenatal care and to provide information on fetal development stages. More recently, during the COVID-19 pandemic, an artificial intelligence (AI) module to answer questions in real-time about the virus was added to inform the users of their COVID test results. Since its initial availability, the application has registered over 10 million hits, 800,000 downloads, and one million scheduled medical appointments.

In the health sector, some new initiatives were implemented to deal with the COVID-19 pandemic. For instance, the Heart Institute (InCor/USP), the Specialized Nucleus in Information Technology (NETi), and the health tech Carenet have joined forces to automate monitoring activities. Through the Orchestra platform, there is integration between electronic medical records and medical equipment (data from the artificial ventilator, electrical impedance tomography, and the multiparameter monitor). The real-time data are presented in the dashboards available for the mobile devices used by intensive care teams. Using the previous process, nurses were spending 33% of their shift circulating within the hospital to collect and record the data from medical records. The new system not only allowed nurses to gain productivity by focusing on their primary activity, it also reduced contamination among the personnel of the hospital by reducing physical interactions. It is possible to see results in the set of measures adopted by InCor, the country's main reference about the treatment of heart disease and COVID, considering that its 40% mortality rate of infected patients is well below the national average of 65%.

15.3.4 *Technological and economic development*

Regarding the development of urbanization for structural transformation and high productivity, we identified great potential in four initiatives: (1) Living Lab (SC), (2) Itaipu Technological Park (PR), (3) Research and

Development Center Development in Telecommunications – CPqD (SP) and (4) Porto Digital (PE). These experiences are related to developing national technologies to support smart cities, besides inserting the respective locations in sectors related to the knowledge-based economy.

On the island of Florianópolis (SC), the Living Lab stands out for being inserted in one of the most active Brazilian innovation ecosystems. The region where the Living Lab is installed has one of the major technological parks in Latin America, Sapiens Park, and a great concentration of Brazilian and foreign ICT companies' offices. The high level of human development, high investments in the technology sector, research institutes and universities of regional importance have attracted companies and highly trained professionals over the past decades. In this context, technology companies with the local city government have instituted a Living Lab to foster startups aimed at smart cities. The highlights are the company Mobilis, which specializes in electric vehicles for rent and sharing; CityTech, a platform that uses AI to generate data for public policies from social networks; and Sigmais, which uses IoT to monitor vehicle traffic and manage parking spaces.

The Itaipu Technological Park (PTI), on the triple border between Brazil–Argentina–Paraguay, emerged to support research and development at the Itaipu Hydroelectric Plant, the second largest dam in the world. PTI is currently working on four themes: (1) Tourism and Cities, (2) Agribusiness, (3) Energy and (4) Critical Security Infrastructure. It is developing projects for alternative energy sources, such as biogas, batteries for stationary and industrial applications, as well as a living lab to test smart cities' technologies. In this laboratory, experiments are being carried out with intelligent public lighting technologies to check people and vehicles, control the light intensity, security cameras, and monitor traffic violations. Tests are also underway to share electric vehicles and bicycles for smart mobility. There are also experiments related to the construction and intelligent buildings and sensing of the ambiance, such as control of humidity, temperature, carbon dioxide, activation of preventive maintenance, reuse of water, and sewage treatment. All data provided by the experiments are managed by the Command and Control Center, which serves as a prototype of the possibilities available to any smart city.

CPqD is also emerging as one of the main technological players for Brazilian smart cities. Located in Campinas (SP), CPqD is the largest center for researching and developing information and communication technologies in Latin America. CPqD created the dojot and Karate platforms for the management of IoT devices based on open-source technologies. Among the main functionalities is the availability of open Application Programming Interfaces (APIs), the management of IoT devices' life cycle, data persistence, and the real-time data analysis interface. The initiative, which has been used by several smart city projects in Brazil, contributes to cost reduction with IoT solutions management.

Finally, inserted in a historic neighborhood, the technology park Porto Digital (PE) has invested in revitalizing buildings and preserving the region's colonial style of Recife, the capital of Pernambuco. Furthermore, the municipal, state, and federal governments facilitated private investments to establish businesses related to the creative economy and the development of ICTs. In this context, Porto Digital currently houses and incubates several companies focused on technologies for smart cities, such as Dona Obra, which provides renovation kits for the communities of Recife; and DISPOR Energia, a subscription platform for solar generators. Startups in Porto Digital are encouraged to solve the Brazilian urban concrete problems, such as the precariousness of popular housing. Among the many projects developed within Porto Digital, the Armazém da Criatividade and the support programs for creative and technological entrepreneurship, such as Mind the Bizz and Incubation, stand out. In 2016, the International Association of Science Parks and Areas of Innovation (IASP) awarded the Armazém da Criatividade de Caruaru and its innovative solutions demanded by companies and local public policies as one of the 10 most innovative projects.

15.3.5 *Urban sustainability and resilience*

Regarding sustainability and urban resilience, there is a set of experiments exploring public–private partnerships (PPP). One of the largest PPP projects is carried out in the city of Belo Horizonte. The project provides for the replacement of 100% of public lighting for Light

Emitting Diode (LED), with remote management service. It is estimated that the measure has already brought savings of 45% on energy supply. The smart devices installed on some specific poles will also allow changing the lighting intensity, exchanging information between devices (such as traffic lights and cameras), and creating a municipal WiFi network coverage.

Some small and medium cities, such as São Luís do Paraitinga (SP), are receiving R&D resources from governmental bodies and agencies, such as the National Electric Energy Agency (ANEEL), to carry out pilot projects. ANEEL supports smart meters' development, deployment of solar energy systems, intelligent public lighting, and regulatory and economic studies of electric buses and bicycles. Aparecida do Norte (SP) has been implementing pilot projects for energy efficiency, intelligent public lighting, and electric mobility to install recharging points for electric vehicles.

Other cities have invested in PPP around solid waste management. For example, the city of Itu (SP) adopted a PPP to install underground containers to deposit and collect waste. The solution, inspired by European cities' examples to not visually pollute the city's tourist attractions, contemplates a space rationalization for pedestrians and car circulation. By the end of a contract signed with service suppliers, the City Hall of Itu estimates the installation of 3,300 containers with sensors that alert when the container is reaching its limit. The containers will also form part of a monitoring network that will show the need for repairs and replacements.

This experience in the city of Itu is like the one in Paulínia (SP), which since 2011, has had an intelligent solid waste collection system. According to data from Paulínia, there was a 50% reduction in monthly expenses with solid waste management. It is noteworthy that intelligent underground containers solve another issue commonly experienced by Brazilian cities—the clogging of culverts on rainy days, which causes flooding. Besides the benefits of the significant reduction in expenses with the maintenance of underground public utilities, Paulínia had the program discontinued in 2017 due to legal conflicts with a private service provider. Mitigating risks and improving the legal system are challenging issues to be addressed.

Some progress is going on with the integration of IoT devices in the Basic Sanitation sector, and the best example comes from the State of São Paulo. Sabesp, the company responsible for the sanitation service in the State of São Paulo and the fourth-largest company in the world, is partnering with startups to develop smart technologies for the sanitation sector. Among the challenges to overcome are water sourcing, water treatment, and reduction of losses in distribution and inhibition of unauthorized uses. One of the initiatives in progress since 2018 has been the installation of low-cost smart water meters, which reduce water losses and alert users when they need to be changed. This is one of the sanitation networks' biggest challenges, as the OECD reports an average loss of 38.45% of drinking water in Brazil.

15.4 Final Considerations

The chapter presented an overview of the Brazilian experience of smart cities. While there is no national strategic plan to guide smart cities, we observe initiatives in regional capitals as well as small and medium Brazilian cities. As these projects are independent from each other, they follow a pattern in terms of objectives. More specifically, these initiatives are focusing on using technology to redefine urbanization and city management to tackle important local challenges such as education, water waste, health, as well as technological and economic development.

These experiences are in line with international policies and commitments, such as the New Urban Agenda 2030 and Agenda 2030 promoted by the UN, to which Brazil is a signatory. Most initiatives on the diffusion of the smart city concept have been presented in national conferences, such as Connected Smart Cities, and performing foundations such as the National Front of Mayors (FNP).

Despite recognizing these new agendas and proposals for urban development, there was an absence of policy coordination at the national level. However, recently, the national government recognized the potential benefit of a coordinated plan to avoid redundancy of effort and investment in the development of similar technological solutions and process innovations in multiple cities. The alignment of isolated national plans, such as the National IoT Plan, and the dissemination of strategic guidelines and national actions, such as the Brazilian Charter of Smart Cities, are promising proposals for rationalizing and directing efforts to the Brazilian smart cities.

Besides greater coordination, there are other challenges to overcome. The fiscal situation of municipalities, especially after the COVID-19 pandemic, affects these entities' capacity to invest in urban development. This context of insufficient funds for promoting smart cities reinforces the need for greater national alignment, especially for acquiring and developing new technologies. Because of the Brazilian legal-bureaucratic framework, a promising solution seems to be the PPP and municipal consortia, which can reduce the contracting time and the cost of analyzing and developing new ICTs.

Another aspect to overcome in Brazil concerns cultural barriers, especially the low-risk aversion which prevents the adoption of innovative solutions [Hofstede, 1986]. In part, this question can be explained by the phenomenon called "blackout of pens", i.e., the fear that public managers must respond to innovations that did not work. Public managers have few incentives to innovate, given the risks of backlash associated with the adoption of new processes and technologies that does not bring value to the stakeholders. Moreover, at the state and municipal levels, we also observe a lack of training around the use of technology by technical staff members.

Finally, the shallow educational level of the Brazilian population is by far the most critical aspect that can delay the development of smart cities in the country. The low level of education and performance in basic educational areas, such as reading, mathematics, and science, makes it difficult for the Brazilian society to be engaged with and to enjoy the benefits provided by a smart city. For example, while the purpose of the development of smart cities is to support the lower-income population by giving them better access to education, health, technology, water system, etc., the reality is that the middle and upper classes are the ones benefiting from smart cities. The lack of knowledge and resources around technology limits the low-income population from participating in smart city projects fully. In summary, the great social inequality and access to the Internet in a large country like Brazil is a challenging factor for developing smart cities. Also, these limitations and the insufficient training of ICT and STEM professionals can represent difficulties for the country to insert itself economically in the high-tech markets, especially those related to industry 4.0.

References

Araujo, S.M.V.G., Fernandes, A.S.A., and Coelho, D.B. (2016). Does metropolitan area management matter in Brazil? Elements of a complex Issue, *disP-The Planning Review*, *52*(2): 17–25.

Banco Nacional de Desenvolvimento [BNDES]. (2018). *Internet das Coisas: Um plano de ação para o Brasil*. URL: https://www.bndes.gov.br/wps/portal/site/home/conhecimento/pesquisaedados/estudos/estudo-internet-das-coisas-iot/estudo-internet-das-coisas-um-plano-de-acao-para-o-brasil.

BBVA Research. (2017). *Urbanization in Latin America*. URL: https://www.bbvaresearch.com/wp-content/uploads/2017/07/Urbanization-in-Latin-America-BBVA-Research.pdf.

Carpenter, M. (2015). The first 50 years of the Community Development Journal, *Community Development Journal*, *50*(1): 1–9.

Castro, M.C., Massuda, A., Almeida, G., Menezes-Filho, N.A., Andrade, M.V., de Souza Noronha, K.V.M., ..., and Giovanella, L. (2019). Brazil's unified health system: The first 30 years and prospects for the future, *The Lancet*, *394*(10195), 345–356.

Centro de Estudos e Debates Estratégicos [CEDES]. (2020). *Cidades Inteligentes: Uma abordagem humana e sustentável*. Brasília: Edições Câmara.

Dempsey, N., Bramley, G., Power, S., and Brown, C. (2011.). The social dimension of sustainable development: Defining urban social sustainability, *Sustainable Development*, *19*(5), 289–300.

Fernandes, E. (2007). Constructing the Right to the City' in Brazil, *Social and Legal Studies*, *16*(2), 201–219.

Garmany, J. (2014). Space for the state? Police, violence, and urban poverty in Brazil, *Annals of the Association of American Geographers*, *104*(6), 1,239–1,255.

Gil-Garcia, J.R., Pardo, T.A., and Nam, T. (eds.) (2015). *Smarter as the New Urban Agenda: A Comprehensive View of the 21st Century City*, vol. 11, Springer.

Harvey, D. (1973). *Social Justice and the City* (No. 04; HT391, H3).

Hu, H., Xie, N., Fang, D., and Zhang, X. (2018). The role of renewable energy consumption and commercial services trade in carbon dioxide reduction: Evidence from 25 developing countries, *Applied energy*, *211*, 1,229–1,244.

Instituto Brasileiro de Geografia e Estatística [IBGE]. (2020). Estimativas de população residente para os municípios e para as unidades da federação brasileiros com data de referência em 1° de julho de 2020, Rio de Janeiro, Brazil: Instituto Brasileiro de Geografia e Estatística.

Kummitha, R.K.R., and Crutzen, N. (2017). How do we understand smart cities? An evolutionary perspective, *Cities*, *67*, 43–52.

Lefebvre, H. (1967). Le droit à la ville, *L'Homme et la société*, *6*(1), 29–35.

Macedo, J. (2008). Urban land policy and new land tenure paradigms: Legitimacy vs. legality in Brazilian cities, *Land Use Policy*, *25*(2), 259–270.

Mannucci, P.M., and Franchini, M. (2017). Health effects of ambient air pollution in developing countries, *International Journal of Environmental Research and Public Health*, *14*(9), 1,048.

Massuda, A., Hone, T., Leles, F.A.G., de Castro, M.C., and Atun, R. (2018). The Brazilian health system at crossroads: Progress, crisis and resilience, *BMJ Global Health*, *3*(4).

Miraglia, P. (2015). Drugs and drug trafficking in Brazil: Trends and policies, *Center for 21st Century Security and Intelligence Latin America Initiative*. URL: https://www.brookings.edu/wp-content/uploads/2016/07/miraglia-brazil-final.pdf.

Nishijima, M., Ivanauskas, T.M., and Sarti, F.M. (2017). Evolution and determinants of digital divide in Brazil (2005–2013), *Telecommunications Policy*, *41*(1), 12–24.

Passero, L.G., Giordani, J.M.D.A., Hugo, F.N., Torman, V.B.L., Camey, S.A., and Hilgert, J.B. (2016). *Contextual and individual factors associated with dissatisfaction with the Brazilian Unified National Health System* (2011–2012), Cadernos de Saúde pública, *32*, e00065015.

Pojani, D., and Stead, D. (2015). Sustainable urban transport in the developing world: Beyond megacities, *Sustainability*, *7*(6), 7,784–7,805.

Rezende, D.A., dos Santos Madeira, G., de Souza Mendes, L., Breda, G.D., Zarpelão, B.B., and de Carvalho Figueiredo, F. (2014). Information and telecommunications project for a digital city: A Brazilian case study, *Telematics and Informatics*, *31*(1), 98–114.

Rolnik, R. (2013). Ten years of the City Statute in Brazil: From the struggle for urban reform to the World Cup cities, *International Journal of Urban Sustainable Development*, *5*(1), 54–64.

Santos Jr, O.D., and Montandon, D.T. (2011). Os Planos Diretores Municipais pós-Estatuto da Cidade: Balanço crítico e perspectivas, Rio de Janeiro: Letra Capital.

Shen, L., Peng, Y., Zhang, X., and Wu, Y. (2012). An alternative model for evaluating sustainable urbanization, *Cities*, *29*(1), 32–39.

Souza, M.D.F.M.D., Malta, D.C., França, E.B., and Barreto, M.L. (2018). Changes in health and disease in Brazil and its States in the 30 years since the Unified Healthcare System (SUS) was created, *Ciencia & Saude Coletiva*, *23*, 1,737–1,750.

Srivastava, M., De Boer, A., and Pijl, S.J. (2015). Inclusive education in developing countries: A closer look at its implementation in the last 10 years, *Educational Review*, *67*(2), 179–195.

United Nations [UN]. (2015). *Transforming Our World: The 2030 Agenda for Sustainable Development*. New York, NY: United Nations.

United Nations [UN]. (2016). *New Urban Agenda: Quito declaration on Sustainable Cities and Human Settlements for all*. New York, NY: United Nations.

United Nations [UN]. (2018). *Revision of World Urbanization Prospects*. New York, NY: United Nations.

Ward, P.M., Huerta, E.R.J., and Di Virgilio, M.M. (2015). *Housing Policy in Latin American Cities: A New Generation of Strategies and Approaches for 2016 UN-Habitat III*, Routledge.

Wilding, P. (2014). Gendered meanings and everyday experiences of violence in urban Brazil, *Gender, Place & Culture*, *21*(2), 228–243.

Zeng, C., Deng, X., Dong, J., and Hu, P. (2016). Urbanization and sustainability: Comparison of the processes in 'BIC' countries, *Sustainability*, *8*(4), 400.

Zhou, X., Li, Z., Zheng, T., Yan, Y., Li, P., Odey, E.A., ... and Uddin, S.M.N. (2018). Review of global sanitation development, *Environment international*, *120*, 246–261.

© 2023 World Scientific Publishing Company
https://doi.org/10.1142/9789811257186_0016

Chapter 16

Virtual Reality for Disaster Management and Recovery

Sheridan Tatsuno

Dreamscape Global
San Francisco, California, USA
Email: smtatsuno@gmail.com

16.1 Introduction

Global concern over climate disasters is rising fast, so even major funds are opening climate funds to "look green" and avoid losing investors. Trillions of dollars are going into these funds and the flood of climate funding has just begun.[1] As an urban planner, business and fiction writer, corporate strategist and serial entrepreneur, I am promoting Virtual and Augmented Reality (VR/AR) tools for visualizing, simulating, planning and telling climate stories and futures.

Hollywood and publishers have done a great job showing dystopian VR fantasy futures, but they have not yet shown how exponential technologies can visualize our current climate disasters. To fill this gap, my fourth VR novel, *Divinely San Francisco*, based on Dante's *Divine Comedy*, imagines and visualizes a massive firestorm in the Bay Area that wipes out a third of the East Bay cities, sending half a million refugees escaping to the streets of San Francisco. I also released an epic poem about our times titled *CHAOS: A Climate Tale*. In 2020, lightning

[1]Tatsuno, S. (2019). Climate Finance: Investors Joining Greta's Call for Action. URL: https://www.linkedin.com/pulse/exponential-tech-climate-action-sheridan-tatsuno.

triggered over a thousand wildfires around the Bay Area, exceeding my worst-case scenario.

How could we use VR/AR and drones for monitoring refugee camps after climate disasters, such as Hurricane Katrina, the Syrian drought, and the Northwest wildfires? How would we use cruise liners, VR/AR, prefabricated buildings (or prefabs), and three-dimensional (3D) printing to rapidly provide shelter? As climate disasters increase, we need to prepare by using fiction, VR/AR storytelling, simulations and gamification to imagine worst-case scenarios so that we can develop viable resilience plans to save lives and property.

My business article titled *Sustainable Development and Smart Communities: The Swedish Experience*[2] observed that Sweden and Denmark are 30 to 40 years ahead of US cities in implementing carbon-neutral solutions, such as Malmo's district heating since the 1990s and Sweden's use of heat pumps to reduce oil imports and biomass burning.

Why is VR storytelling and simulations so important for climate action? In early 2016, my Swedish VR company developed VR simulations of hurricanes, sea rise and flooding for our client, Copenhagen Malmo Port. Our developers—all gamers who love creating disasters and mayhem—said we could also simulate fires, explosions, and other disasters. Their comments fired my imagination. I just needed a potential client.

16.2 Simulating Wildfires: Pacific Gas and Electric's Camp Fire Disaster

In late-2016, I was pitching VR to Pacific Gas and Electric (PG&E), which wanted to visualize its 700 facilities and more as part of a $2.5 billion rebuilding program. In late-2017, wildfires struck Northern California. This was the perfect opportunity to pitch my VR for disaster preparation! So, I suggested creating VR wildfire scenarios

---

[2] Tatsuno, S. (2019). Sustainable development and smart communities: The Swedish experience, in: Nuttall, W.J., Gibson, D.V., Trzmielak, D., and Ibarra-Yunez, A. (eds.), *Energy and Mobility in Smart Cities*, pp. 61–80. Ice Publishing. DOI: 10.1680/emsc.64256.061.

and visualizing how high winds could affect the path of wildfires into towns. Unfortunately, PG&E said that "We don't like to talk about fires. Perhaps later". In late-2018, the Camp Fire wildfires struck, killing many residents, destroying the town of Paradise, and sending PG&E into a $30 billion bankruptcy. I felt bad for not persisting, since our VR could have helped PG&E visualize where to cut trees touching power lines, simulate fires spread by high winds, and train towns and firefighters how to respond quickly.

To prevent future fire disasters, I wrote about the use of VR for Forest Fire Management and Recovery.[3] Recently, towns near Sacramento, California began using VR to train firefighters,[4] which I believe will become mandatory, like flight simulators for airline pilot training, as VR is perfect for dangerous, distant, repetitive, and collaborative tasks.

16.3 Simulating Hurricanes and Typhoons

After Category 5 typhoons hit Guam and Japan, Hurricane Maria hit Puerto Rico and Hurricane Florence hit the Carolinas, I wrote this chapter about VR and AR for Hurricane Resilience Planning and Disaster Recovery. Typhoons and hurricanes have already caused massive losses and damage. Japan was shut down for a week, which probably cost it $100 billion in lost economic activity, in addition to the loss of lives and damaged infrastructure and property. Needless to say, Japanese insurance companies and cities are beefing up their resilience planning since Japan is plagued by so many disasters—typhoons, floods, mudslides, sea rise, earthquakes, and the Fukushima meltdown. In all cases, VR/AR could be used for simulating, visualizing, planning, designing, and recovering from disasters.

---

[3] Tatsuno, S. (2018). *VR for Forest Fire Management and Recovery*. URL: https://www.linkedin.com/pulse/vr-forest-fire-management-recovery-sheridan-tatsuno/.

[4] CBS Sacramento. (2019). *Cosumnes Firefighters Training with Virtual Reality*. URL: https://sacramento.cbslocal.com/2019/12/12/cosumnes-firefighters-training-virtual-reality/.

16.4 How to Use Extended Reality (XR) Storytelling and Simulations for Disaster Resilience Planning

As my Swedish VR team has learned from our projects on sustainable city planning and construction training, VR/AR can be used for the entire disaster management process:

- *Threat and Opportunity Identification*: Cities, builders, insurance companies, businesses and homeowners can identify potential threats, such as wildfires, flooding, hurricanes, etc., using NASA and NOAA data, Geographic Information Systems (GIS), VR maps and previous disaster data.
- *Scenarios:* Based on these threats and opportunities, communities can develop a variety of best-case to worst-case (black swan) scenarios with city officials and planners at a conceptual level, such as winds at 100 miles per hour causing massive firestorms or a Category 5 hurricane.
- *Simulations*: Using VR, planners can then simulate possible disaster scenarios so officials and citizens can visualize them, as we did by showing one-foot-per-second sea rise for 30 seconds in Copenhagen that submerged the city. NASA and NOAA climate forecasts can be used to provide real-time and historical data on VR maps. Builders, planners and insurance companies can estimate the potential loss of life and damage.
- *Training*: Using the simulations, first responders can train on various disaster scenarios, such as a Category 5 hurricane in Florida or wildfires in California. Not only frontline firefighters and other first responders, but all city officials, citizens and other towns and cities could train and prepare using these disaster simulators. Using the Unity and Unreal VR gaming engines, cities could gamify disaster training through hackathons to challenge teams of disaster experts and VR/AR developers to simulate disasters and prepare resilience plans.
- *Monitoring*: Using Internet of Things (IoT) sensors, satellite data, drones, ground cameras and other data capture methods, local officials can monitor disasters such as hurricanes and wildfires in real time. Moreover, they could even virtually fly, drive or walk through

the disasters à la Terminator 2, which can be an eerie experience. Fire chiefs could inspect wildfires by flying or driving through the virtual simulations as they are happening so they can better coordinate firefighting efforts.

- *Storytelling*: 360-degree videos are already being used to capture the recent Camp Fire ruins and to tell stories by the survivors. In the future, 3D volumetric capture can be used to reconstruct the disaster site for VR walk-throughs, public education, first responder training, and reconstruction planning.
- *Analysis*: After the disaster, satellite, drone, IoT, communications and other data can be used to reconstruct and analyze the disaster, and first responder activities to derive lessons and develop best practices.
- *Reporting*: Live disaster reporting for the mass media and follow-up studies can be aggregated into disaster databases and provided to first responders, cities, trainers and government agencies, so everyone can visualize and learn from the disasters.
- *Public Education*: These XR disaster databases would be valuable for educating the public, K-12 schools and colleges about disaster hazards and best practices for planning, training and responding to disasters. As climate disasters increase, climate training and education should be mandatory in all public schools.
- *Reconstruction*: VR/AR can be used for designing, visualizing and walking through emergency housing, new buildings, smart grids and other infrastructure after disasters, and involve homeowners, insurers and citizens in the reconstruction process. The goal would be safe, resilient, sustainable cities and communities.

So, there are many beneficial uses of XR for disaster resilience planning, management and recovery, which I believe will be mandatory in the future since XR training and simulations, as we have seen in the airline industry, are much safer, cheaper, and easier than using physical systems and equipment, plus trainees can run multiple scenarios even in the comfort of their homes. My VR gamer friends are particularly excited because they are getting tired of hit-or-miss video games and want to do something useful for society. As we say in urban planning, "Forewarned is forearmed!" And with XR, being trained is even better.

16.5 Simulating Wildfires that Threaten Cities

California's disastrous forest fires are worsening fast, but they come as no surprise to environmental groups. In March 2018, the Sierra Club and other nature organizations urged Governor Jerry Brown to discourage urban development in fire zones (i.e., the Sierras and coastal foothills) based on the 2017 fires, but it takes time to implement their recommendations.

Given the profusion of firestorms, what can California and other fire-prone regions do to reduce loss of life and property? Applying the above VR planning approach to forest fires, the following are what can be done:

- *Forest fire preparedness and planning*: Preventing the loss of lives and property is paramount. As the Sierra Club recommends, California and other states should prohibit and/or limit urban development in high-risk fire zones and cluster them closer to lakes and other sources of water where people can escape and firefighters can access water quickly. CalFire should establish a Virtual Fire Center to manage all potential and ongoing fires. VR "forest fire simulators" could simulate and visualize any urban layouts and the potential impacts of storms, winds, lightning strikes, and forest fires on any specific community. When integrated with CalFire and local government data, the IoT and AI, predictive analytics could forecast the need for firefighters, police, medics, hospitals and emergency housing. CalFire and other first responders could be trained using avatars within the disaster scenarios via *training gamification*, by offering forest fire challenges and scoring the trainee actions. Given the impact of high winds, fire scientists will need to modify their models to account for global warming and longer dry seasons.
- *Disaster management*: During forest fires, real-time VR satellite and drone imagery updated on the ground with CalFire imagery and data could enable governments, first responders and businesses to fly, drive or walk through the disaster virtually and monitor the situation using real-time data, which would enable them to deploy and manage fleets, supplies and rescue workers. The VR scenes could be

webcast globally and viewed on two-dimensional (2D) displays (i.e., smartphones, tablets, PCs, TVs or large screens) and VR headsets by government officials, emergency responders, and news media and combined with emergency services, team sites, global firefighter online training, and forums to coordinate rescue efforts. People living in fire zones could see ongoing fires and emergency zones, locations of family members and friends who post their smartphone locations, and local evacuation centers. Satellite and drones could constantly update VR imagery to create a live and historical database of all fires, first-responder, medical, city and citizen activities.

- *Disaster recovery and rebuilding*: VR and AR can be used for the entire rebuilding process from disaster site assessments, alternative concept designs and community planning to final architectural designs (BIM), construction, and post-recovery data monitoring. VR could plan, design and show potential temporary housing sites at recreational vehicle (RV) parks, state fairgrounds, parking lots and other public areas where RVs, mobile homes, medical vans, and food kitchens could be parked to provide emergency housing, power, food, water and toilets. For rebuilding destroyed towns, VR could be used for architectural and engineering designs, updated during construction with AR images overlaid to compare VR/BIM models with actual construction progress. For example, California cities could use VR maps to show their entire region, towns and cities upon which they could visualize the location of temporary and permanent buildings and infrastructure for people displaced by forest fires. Displaced retailers could set up virtual shops to sell using a mix of VR and AR as a way to revive their businesses until their brick-and-mortar stores are rebuilt. Schools and nonprofits could host crowdfunding campaigns, which could be shown in VR.
- *Reporting and Storytelling* are also important for educating, training and preparing for worsening wildfires.[5]

[5] Tatsuno, S. (2019). *XR Storytelling for Climate Futures*. URL: https://www.linkedin.com/pulse/xr-storytelling-climate-futures-sheridan-tatsuno/.

There are many other ways that VR/AR can be used with IoT, AI and Blockchain for managing forest fires and other disasters through a single virtual Fire Command and Control Center that provides 3D/VR and data visualization of all fire zones, both in real time or historically. The technologies are available but they need to be customized and implemented for forest fires. In the future, fire preparedness, management and training will become a much faster, more effective tech-driven collaborative effort.[6]

[6] See VR demos at: www.onereality.com and www.facebook.com/groups/263449064157532/.

© 2023 World Scientific Publishing Company
https://doi.org/10.1142/9789811257186_0017

Chapter 17

Smart Sustainable Cities and Entrepreneurship: A Proposal for a Non-aggregative Approach

Enrico Ivaldi,[*,‡] Andrea Ciacci[†]

[*] *University of Genoa, Department of Political Sciences, Genoa, Italy, C.I.E.L.I., the Italian Centre of Excellence on Logistics Transports and Infrastructures, Genoa, Italy and Centro de Investigaciones en Econometría – CIE University of Buenos Aires*
[†] *University of Genoa, Department of Political Sciences, Genoa, Italy*
[‡] *Corresponding author, Email: enrico.ivaldi@unige.it*

17.1 Introduction

According to a recent definition of the Italian National Observatory of Smart Cities, the smart city is an "abstract projection of communities of the future, an applicative and conceptual perimeter defined by a set of needs that find answers in technologies, services and applications attributable to different domains" [ONSC, 2020]. The concept of the smart city however is not identifiable with a single semantic solution. Depending on the context or perspective assumed, the expression "smart city" can assume different meanings. A smart city employs human capital, social capital, and information and communications technologies (ICTs) to create a network with the support of hard and soft infrastructures [Hollands, 2008; Washburn *et al.*, 2009]. Making a city smart means, therefore, subjecting it to a series of coordinated interventions to enhance its energy and environmental sustainability and the quality of services to citizens, so as to guarantee participation and active citizenship [Giffinger *et al.*, 2007]. In short, in a real smart city, every problem is no longer tackled individually, but becomes part of a complex system and a new global vision of the urban space. For all this to really work, an integrated network is needed; that is, a common platform coordinated by a single governance that provides the necessary funding to achieve the political and social identity of smart cities. The presence of a large technological infrastructure is what harmonizes all these needs, generating inclusion, integration and efficiency applied to everyday life.

The identity of the smart city must be built around some fundamental concepts whose common denominator is environmental and energy sustainability. For this reason, participation is a key step of the transformation process, which includes keywords such as computer literacy, inclusion, training, awareness and involvement [Höjer and Wangel, 2015]. To summarize the above, it could be said that the term "smart city" refers to an urban complex with strategic, structural and cultural characteristics that guarantee a projection towards a future of greater simplicity, well-being and knowledge for all those who live there.

The United Nations (UN) estimated that, by 2050, 66% of the world population will live in cities [United Nations, 2015]. This implies significant challenges concerning environmental and social sustainability [OECD, 2012]. In addition, cities in their contemporary form appear as a source of environmental and social problems—unemployment, deteriorating housing, concentrations of poor and minority ethnic groups, intolerance and racism, discrimination, social exclusion and environmental pollution. In short, cities were dumping places and so need to be avoided. The smart city, therefore, is inextricably linked to the goals of sustainability. In recent years, the concept of sustainable development has been introduced in local and regional planning as a new market-based strategy for land and housing development that is supposed to be able to integrate economic, social and ecological sustainability [Krueger and Gibbs, 2008; Scheller and Thörn, 2018].

A sustainable city is designed with consideration for social, economic and environmental impact, commonly referred to as the triple bottom line. These elements are interconnected and all of them are crucial for the pursuit of individual and social well-being [Alaimo and Maggino, 2019]. Therefore, the smart sustainable city can use ICTs to carry out social, cultural and organizational innovation processes in order to improve the quality of life, employment levels and competitiveness, and thus meet the needs of current and future generations, ensuring their economic, social and environmental sustainability [Ibrahim *et al.*, 2018].

The words "smart" and "sustainable" refer to all the policies that optimize the city's functionality and entail the protection of the environment, stimulation of economic growth in the local context, and improvement of the people's quality of life thanks to infrastructures, processes, and services supported by ICTs [Aina, 2017; Anthopulos, 2017; Bibri and Krogstie, 2017; Kramers *et al.*, 2016; Martin *et al.*, 2018; Tokody and Schuster, 2016]. Environmental management, governance, entrepreneurship and citizen participation are mainstays of any smart city [Escamilla Solano *et al.*, 2017].

Additionally, literature about Smart and Sustainable Cities highlights the benefits accrued through learning, through knowledge sharing, and through deep labor market. However, these effects depend on the city's dimension of reference, and tend to be larger as the city is larger [Porter, 1998; Porter, 2001; Santinha *et al.*, 2019].

Cities capable of both ensuring a better quality of life and driving economic activities are likely to grow faster than others [Ciacci *et al.*, 2020], favored by the rise in social mobility. For this reason, national and European Union (EU) policies are increasingly focusing on the theme of the smart city. Only advanced and integrated technologies at all levels of the urban ecosystem seem to be able to address this issue and guarantee considerable savings to Public Administrations, without penalizing the quality of the services offered [Garau and Pavan, 2018].

A smart sustainable city also doubles as a "knowledge city", which improves the competitiveness of urban contexts in the global knowledge-based economy [Glaser *et al.*, 2001]. By contrast, extant economic literature on the knowledge-based economy mainly considers large metropolises, focusing primarily on cities in North American [Glaser *et al.*, 2001, Glaser *et al.*, 2014] or in emerging economies [Florida, 2002; Florida *et al.*, 2008]. Only a few academic contributions have addressed the knowledge-based economy and its development drivers in smaller urban ecosystems.

Following this thread, the research question is the following—is the smart sustainable city environment in Italy a stimulus for the urban knowledge-based economy?

The novelty of this study consists in the use of the Partially Ordered Set (POSET) theory, an instrument that permits to study different phenomena, suitable to reply at the research question. The POSET theory is based on a formative and non-aggregative conceptual setting. The POSET method is not yet widespread, but its diffusion is growing.

The analysis considers a sample of 20 regional capitals in Italy, in order to include both medium-sized and large cities. In this way, it is possible to capture the differences in their smart and sustainable characteristics and attitudes towards the knowledge economy.

The Italian context is especially interesting because smart city policies are spread nationwide and are rooted in a redesign of the governance at the local level [Smigiel, 2019]. Moreover, there is a high differentiation in the Italian urban development (for example, between large and small cities, and northern and southern cities).

17.2 Materials and Methods

Cities have their own metabolisms, patterns of activities, consumption and waste and ecological footprints that reach beyond their borders. Social, economic, political and cultural features of cities communicate and react with the spatiality and build cities' environment [Klopp, 2017]. Any set of indicators would thus necessarily seem rigid and unable to fully reflect urban complexity, and the many scales on which the impact of cities is felt [Holden, 2013]. However, indicators have multiple uses. Therefore, they can contribute to the assessment and benchmarking of conditions and trends across space and time, they can help to monitor progress toward goals and targets, inform planning and decision-making, raise awareness, encourage political and behavioral changes, promote public participation, and foster better communication [Holden, 2006; Moreno Pires *et al.*, 2014].

When considering the nature of the variables for measuring a smart sustainable city and their definitions, it appears that the model of measurement—with reference to the relationship between the latent variables and the basic indicators selected—is formative, because indicators are regarded as causing the phenomena rather than being caused by them, which is true in the reflective approach. As a consequence, changes in formative indicators cause changes in the value (and meaning) of the latent variable [Ivaldi *et al.*, 2020a]. This approach shows that indicators are not interchangeable, because, by omitting an indicator, part of the construct is also omitted. The correlations between the indicators do not influence and are not explained by the measurement model; internal consistency has little relevance—two indicators with no correlation can both be used to describe and interpret the same construct [Maggino, 2017].

The choice of domains is closely related to the achievement of the desirable goals for smart sustainable cities. In this term, on the basis of the main literature, we have identified the following components: "Sustainable Urban Environment" (SUE); "Sustainable Social Services" (SSS); "Sustainable Local Transport" (SLT); and "Sustainable Safe Environment" (SSE). The selected domains cover the multidimensional nature of smart sustainable cities and offer a focus on citizens' quality of life, which contributes to attract and retain knowledge workers. These domains were built in a previous work [Ivaldi *et al.*, 2020b] on the basis of the literature concerning smart cities and smart sustainable cities [Garau and Pavan, 2018; Giffinger *et al.*, 2007]. They differ sufficiently in the features they describe to avoid the risk of self-correlation.

Regarding entrepreneurship, it is possible to observe the domain "Knowledge Economy and Innovation" (KEI) that represents the knowledge-based attitude of each city [Ivaldi *et al.*, 2020b; Penco *et al.*, 2020]. This domain is a proxy with the function of illustrating the health and stability of the knowledge-based economy at the city level referred to the range of activities that employ knowledge workers, to attitudes to product and process (patents), and to the wealth of urban inhabitants. These profiles have been selected consistently with the literature on knowledge-based economy [Méndez and Moral, 2011], allowing for the availability of data for the sampled city (Table 17.1). Higher values of the variables represent better situations and vice versa.

Table 17.1. Cities and domains, variable used.

Cities	Abbr	Sustainable urban environment (SUE)	Sustainable social services (SSS)	Sustainable local transport (SLT)	Sustainable safe environment (SSE)	Knowledge economy and innovation (KEI)
Ancona	AN	5.44	6.02	3.39	3.36	3.67
Aosta	AO	6.25	3.64	3.06	0.34	2.86
Bari	BA	1.47	3.99	1.89	0.29	1.27
Bologna	BO	10.19	12.74	4.56	0.13	7.93
Cagliari	CA	5.67	4.70	6.46	6.68	2.73
Campobasso	CB	4.05	3.51	2.36	2.16	0.37
Catanzaro	CZ	0.00	4.56	2.99	3.19	0.00
Florence	FI	7.70	7.93	5.02	0.31	4.07
Genoa	GE	4.77	7.69	4.25	0.00	5.54
L'Aquila	AQ	4.93	3.05	2.06	3.50	2.78
Milan	MI	3.57	10.60	8.87	0.11	9.53
Naples	NA	5.23	0.00	0.95	0.44	1.24
Palermo	PA	1.97	3.16	0.00	0.10	0.63
Perugia	PG	8.27	7.49	3.15	0.54	1.17
Potenza	PZ	8.94	2.27	0.68	2.39	0.09
Rome	RM	5.59	8.55	3.52	0.00	6.51
Trento	TN	12.29	13.51	6.13	3.98	3.78
Trieste	TS	7.60	11.57	6.81	10.67	4.11
Turin	TO	12.02	5.25	3.75	0.22	6.74
Venice	VE	5.44	6.02	3.39	3.36	3.67

Figure 17.1 represents the capital cities of the Italian regions for a better understanding of the text.

In a more or less explicit way, the classic paradigm with which multidimensional assessment problems are addressed in the socioeconomic context is based on the attempt to aggregate elementary indicators in synthetic indices. The latter allow a measurement of the latent feature and the construction of rankings between statistical units. In most cases, the aggregation takes the form of a weighted average of indicators, built

Figure 17.1. Italian regional capital cities.

through the use of models for latent variables or obtained more simply by identifying the weights of the weighted average through heuristic procedures, or with the use of experts. In any case, the latent one-dimensional nature of the phenomenon of interest and aggregation as a method of synthesis of information are implicitly assumed [Fattore, 2013].

In the case of smart sustainable cities, a one-dimensional reduction of the problem is often unworkable and liable to produce misguiding images of facts and reality. The versatility of the concept emerges clearly, as well as the need to identify tools and methodologies capable of reflecting its complexity. In practice, how can we assess the basic features of an actual "smart city", to be chosen in a group of urban centers? It is essentially a matter of multidimensional comparison between "individual profiles". Technically, it is a comparison between the vectors composed of the scores of each statistical unit on the domains considered [Ivaldi *et al.*, 2020]. For example, if we consider three cases of possession or non-possession of a certain attitude (expressing possession 1 or non-possession 0), there are eight possible profiles: 111, 110, 101, 011, 100, 010, 001, 000. Obviously, Profile 111 (possession of all the three attitudes considered) represents the best situation, just as Profile 000 represents the worst situation. However, not all profiles are comparable. For example, Profiles 110 and 101 are incomparable profiles, as well as Profiles 110 and 001. Indeed, the eight profiles listed constitute a partially ordered set, or POSET [Davey and Priestley, 2002], that is a set in which not all pairs of elements can be ordered together. The existence of incomparability, which is reflected in the partial nature of the rankings and in the inability to respond immediately to the question above, might initially seem a "disturbance" with respect to the objectives of the evaluation. However, it is a direct consequence of the complexity and versatility of the problem, and it would be a stretch not taking it into account [Alkire and Foster, 2007; Fattore *et al.*, 2011].

The POSET is therefore the mathematical structure allowing to represent the data of interest in an appropriate way and to set evaluation procedures consistent with the nature of the data and efficient in extracting information [Fattore *et al.*, 2011].

Within this approach, the POSET has become a point of reference over the years, as demonstrated by many works in different fields of research with references to European countries [Annoni and Brüggemann, 2009], subjective well-being [Alaimo *et al.*, 2020; Fattore *et al.*, 2015], poverty [Beycan *et al.*, 2018] and deprivation [Arcagni *et al.*, 2019; Ivaldi *et al.*, 2020].

According to Fattore [2017], the POSET evaluation is a way to obtain a complete order out of a partial order. Once the evaluation scores have been assigned to the elements of a POSET, they can be arranged in a linear order. The procedure allows assigning to each element of a finite POSET a score representing its position in a "low-high" axis. In this way, we obtain the average rank, i.e., the average position assumed by a profile in each of the considered linear extensions—the lower the average rank, the better the position taken by the profile in the order.

The ranking procedure has developed following these steps [Fattore, 2017]:

(1) We extracted all the linear extensions of a certain order structure Π^* and create $\Omega(\Pi^*)$;
(2) For each element $p \in \Pi^*$ and for each $l \in \Omega(\Pi^*)$, we assigned the rank $r_l(p)$ of p in l, which is defined as 1 + the number of edges linking p to the maximum of l;
(3) We calculated the average $r(p)$ of $r_l(p)$ over $\Omega(\Pi^*)$ for each $p \in \Pi^*$;
(4) Then we have obtained the average rank of each element of the distribution.

This value expresses the attitude of the city to be "smart sustainable"; smaller values of average rank represent better situations and, vice versa, higher values represent worst situations.

All the results and graphical representations were obtained through the appropriate processing with the RStudio software, using the functions offered by the PARSEC package [Arcagni and Fattore, 2014].

17.3 Results and Discussion

Table 17.2 and Figure 17.2 show the measurement and representation of the range of variation for each city. They measure the maximum-minimum range within which a city is located, providing a reference parameter to describe the situation of uncertainty in which a city oscillates. The fuzziest cases are represented by Milan, in the northwest of Italy, and Potenza, in the south of Italy, both with a variation range of 18. In these two cases, only one comparability can be defined for each of the two cities.

On the other hand, it is possible to identify in Trento, (maximum-minimum 1 to 4), Palermo (maximum-minimum 13 to 20) and Trieste

Table 17.2. Average rank statistics.

Unit	Ave. rank value	Min.	Max.	Range
Trento	1.49	4	1	3
Trieste	2.43	9	1	8
Cagliari	6.1	14	2	12
Perugia	6.18	16	2	14
Ancona	6.77	15	3	12
Bologna	6.89	17	2	15
Florence	7.39	17	2	15
Milan	9.27	19	1	18
Turin	10.04	19	2	17
Venice	10.04	19	2	17
Potenza	11.14	20	2	18
Aosta	12.64	19	4	15
Campobasso	13.27	19	5	14
L'Aquila	13.59	20	4	16
Rome	14.13	20	4	16
Catanzaro	14.36	20	5	15
Naples	14.87	20	6	14
Genoa	15.11	20	5	15
Bari	15.52	20	7	13
Palermo	18.76	20	13	7

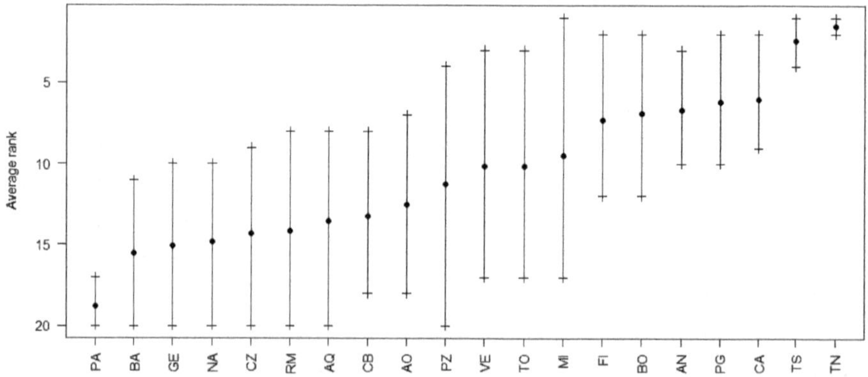

Figure 17.2. Average rank graph with range of minimum and maximum.

(maximum-minimum 1 to 9) the cities that show the smallest oscillations.

It should be noted that 11 cities out of 20 have a range of variation equal to or greater than 15. The data expressed the difficulty of ranking univocally the analyzed cities. Therefore, the picture appears uncertain and hazy.

Table 17.3 shows the numerical results of the average rank and the position that units cover in the KEI rank. The best city by average rank value is Trento (1.49), which corresponds to a position closer to the

Table 17.3. Average rank and KEI rank.

Unit	Average rank value	KEI rank value
Ancona	6.77	9
Aosta	12.64	10
Bari	15.52	14
Bologna	6.89	2
Cagliari	6.10	12
Campobasso	13.27	18
Catanzaro	14.36	20
Florence	7.39	7
Genoa	15.11	5
L'Aquila	13.59	11
Milan	9.27	1
Naples	14.87	15
Palermo	18.76	17
Perugia	6.18	16
Potenza	11.14	19
Rome	14.13	4
Turin	10.04	3
Trento	1.49	8
Trieste	2.43	6
Venice	10.04	13

middle rank in the KEI rank. A clear gap to subsequent cities is established by Trieste (2.43), to which is associated the sixth position in the KEI rank.

Cagliari, Perugia, and Ancona obtain a score between 6.10 and 6.77. Bologna approaches them with a score of 6.89, although it is placed at a lower level in the POSET chart (Figure 17.2). This is due to the lower number of relationships that Bologna establishes compared to Perugia, Cagliari and Ancona, and to the impossibility to compare Bologna with the cities on the third level. Florence obtains an average rank value similar to that of the KEI ranking, of 7.39 and 7 respectively. Marked differences between ranks distinguish Turin (10.04 and 3), Genoa (15.11 and 5), Perugia (6.18 and 16), Potenza (11.14 and 19), Rome (14.13 and 4), as well as Milan, which, despite its first place in the KEI ranking, scores 9.27 in the average rank.

The Hasse diagram is formed by nodes, connected to each other through a downward path; nodes that are not connected are incomparable [Fattore, 2017]. Looking at the Hasse Diagram in Figure 17.3, it is possible to see that Trento and Trieste are at the highest level of the structure. Due to the particular composition of the profile, Milan establishes a single dominance relationship (with respect to Palermo) and is not dominated by any city. At the second level of the ordered structure, there are Perugia, Cagliari and Ancona, while the greatest concentration

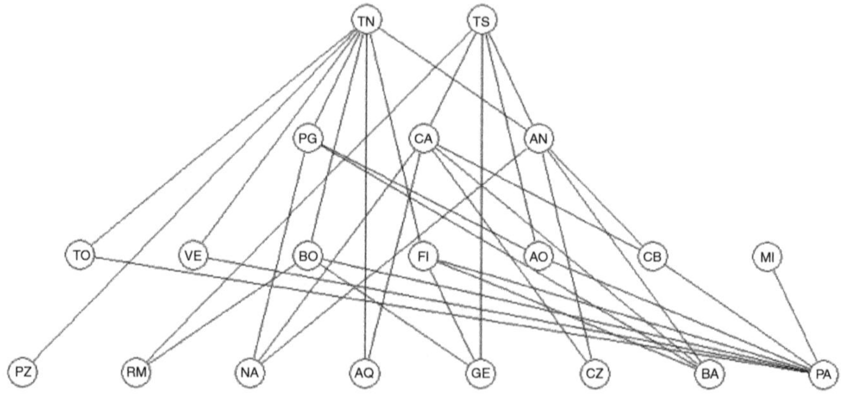

Figure 17.3. A Hasse diagram for Italian regional capital cities.

of cities is at the third and fourth, although the relations that connect them with the cities above are quite heterogeneous. Given the high concentration of units at the third and fourth level, and the impossibility of establishing relations between cities at the same level, many cities are incomparable to each other.

For example, Potenza has only one relation, with Trento. Although it is positioned at the same level, Palermo is dominated by a large number of cities, showing a greater weakness in its profile values.

The reduction of a set of indicators to a single number flattens the differences between statistical units, so that incomparable profiles become comparable. POSET seems to overcome this limit by attributing different values to units with different combinations in the basic indicators, whereas an aggregative index can often attribute a similar value to them [Alaimo and Maggino, 2019].

The Pearson's correlation coefficient explains the contribution of each domain to the "Knowledge Economy and Innovation" (KEI) ranking which represents the knowledge-based attitude of each city. The KEI ranks are moderately correlated with average rank (0.344).

The POSET of smart sustainable cities in Italy is a potentially moderate predictor of high KEI attitude at the city level and a smart sustainable environment can contribute to the development of a knowledge-based economy, even if in a different way depending on the dimension of the cities [Yigitcanlar, 2007]. At the same time, it appears that the development of a knowledge-based economy can increase the level of smartness and sustainability of the city, since there is a specific demand for city living and high quality of life by knowledge workers, which confirms the extant literature [Florida 2002; Florida *et al.*, 2008; Ivaldi *et al.*, 2020; Yigitcanlar *et al.*, 2007].

The best positioned cities have recently put in place public policies. Such policies have the goal of enhancing both sustainability and quality of life, so as to increase the ability of cities to attract and retain talents. The contemporary economy fosters competition between cities to attract investments, businesses, inhabitants, and tourists, and to increase their citizens' satisfaction. Strategic planning, marketing strategies, and city branding are some of the main tools they use to compete.

17.4 Conclusions

Currently, half of the world's population lives in an urban environment—3.5 billion people, which will rise to six billion in 2030. The phenomenon of urbanization, now consolidated in the West, is growing fast in emerging countries; it is estimated that, in the next few years, 95% of urban sprawl will occur there [United Nations, 2015].

Our time poses major challenges—we have to meet the basic needs of an increasing urban populations while guaranteeing the integrity of critical ecosystems, addressing climate change and stimulating economic productivity and social inclusion.

In addition to the digital revolution, which is also indispensable to move the country towards a circular economy model, innovation must concern social innovation as well. For this reason, investment in human and social capital is relevant, especially in cities.

Coordination between all levels of government with the urban dimension at the center is needed, because social and economic problems are gathering in cities, but skills and resources to solve them may be also found there. Small urban ecosystems can adopt smart sustainable policies to increase the range and efficiency of their services (thus enhancing the quality of life), while the knowledge-based development of smart communities can benefit the whole economic system.

In literature, there is a bidirectional relationship between entrepreneurship and smart cities. First, entrepreneurs initiate technological interventions that help cities undergo socio-technical transitions and become smart cities [Ivaldi and Ciacci, 2020]. Second, the technologies being adopted in cities generate data which helps enterprises to explore new opportunities [Kummitha, 2019]. In the present chapter, this relationship partially emerges.

The present work obviously has limits. Firstly, working on the regional capitals brings together cities of different sizes that could have a different relationship with the KEI segment depending on the dimension. Secondly, the analysis could be carried out with a higher number of variables that could provide a more multifaceted representation of the phenomenon.

The choice of the POSET method to conduct our analysis is due to a specific feature of this methodology. Indeed, since it provides synthetic indicators without attributing any aggregation, it allows treating cardinal or ordinal data in a consistent way and avoiding the disadvantages of both aggregative-compensative and counting approaches.

References

Aina, Y.A. (2017). Achieving smart sustainable cities with GeoICT support: The Saudi evolving smart cities, *Cities*, *71*, 49–58.

Alaimo, L.S., Ciacci, A., and Ivaldi, E. (2020). Measuring sustainable development by non-aggregative approach, *Social Indicators Research*. URL: https://doi.org/10.1007/s11205-020-02357-0.

Alaimo, L.S., and Maggino, F. (2019). Sustainable development goals indicators at territorial level: Conceptual and methodological issues—the Italian perspective, *Social Indicators Research*. URL: https://doi.org/10.1007/s11205-019-02162-4.

Alkire, S., and Foster J. (2007). Counting and multidimensional poverty measures. *OPHI Working Paper Series 7*. Oxford.

Annoni, P., and Brüggemann, R. (2009). Exploring partial order of European countries, *Social Indicators Research*, 92(3), 471–487.

Anthopoulos, L. (2017). Smart utopia vs. smart reality: Learning by experience from 10 smart city cases, *Cities*, *63*, 128–148.

Arcagni A., and Fattore M. (2014). PARSEC: An R package for POSET-based evaluation of multidimensional poverty, in: Brüggemann, R., Carlsen, L., and Wittmann, J. (eds.), *Multi-Indicator Systems and Modelling in Partial Order*, pp. 317–330. Berlin: Springer.

Arcagni, A., Barbiano di Belgiojoso E., Fattore M., and Rimoldi, S.M.L. (2019). Multidimensional analysis of deprivation and fragility patterns of migrants in Lombardy, using partially ordered sets and self-organizing maps, *Social Indicators Research*, *141*, 551–579.

Beycan, T., Vani, B.P., Brüggemann, R., and Suter, C. (2018). Ranking Karnataka districts by the Multidimensional Poverty Index (MPI) and by applying simple elements of partial order theory, *Social Indicators Research*. URL: https://doi.org/10.1007/s11205-018-1966-4.

Bibri, S.E., and Krogstie, J. (2017). Smart sustainable cities of the future: An extensive interdisciplinary literature review, *Sustainable Cities and Society*, *31*, 183–212.

Ciacci A., Ivaldi E., and Soliani R. (2020). A potential business environment of a smart cities: A subjective approach, in: Dinçer, H., and Yüksel, S. (eds.),

Business and Finance Innovation: Multidimensional Policies for Emerging Economies. Bingley, UK: Emerald.

Davey, B.A., and Priestley, B.H. (2002). *Introduction To Lattices and Order*. Cambridge: Cambridge University Press.

Escamilla Solano S., Plaza Casado P., and Flores Ureba S. (2017). Smart cities and sustainable development. A Case Study, in: Peris-Ortiz, M., Bennett, D., Pérez-Bustamante Yábar, D. (eds.), *Sustainable Smart Cities. Innovation, Technology, and Knowledge Management*. Cham: Springer.

Fattore M., Brüggemann R., and Owsiński J. (2011). Using POSET theory to compare fuzzy multidimensional material deprivation across regions, in: Ingrassia, S., Rocci, R., and Vichi, M. (eds.), *New Perspectives in Statistical Modeling and Data Analysis*. Springer-Verlag.

Fattore M., Maggino F., and Greselin F. (2011). Socioeconomic evaluation with ordinal variables: Integrating counting and POSET approaches, *Statistica & Applicazioni*, Partial Orders in Applied Sciences, Special Issue, pp. 31–42.

Fattore, M. (2013). La valutazione statistica del benessere: Ordinamenti parziali e nuovi paradigmi metodologici, *Statistica e società*, *2*(2), 1–4.

Fattore, M. (2017). Synthesis of indicators: The non-aggregative approach, in: F. Maggino (ed.), *Complexity in Society: From Indicators Construction to their Synthesis*, pp. 192–212. Cham: Springer.

Fattore, M., Maggino, F., and Arcagni, A. (2015). Exploiting ordinal data for subjective well-being evaluation. Statistics in transition, *Journal of the Polish Statistical Association*, *16*, 409–428.

Florida, R. (2002). The economic geography of talent, *Annals of the Association of American Geographers*, 92(4), 743–755.

Florida, R., Mellander, C., and Stolarick, K. (2008). Inside the black box of regional development—human capital, the creative class and tolerance, *Journal of Economic Geography*, *8*(5), 615–649.

Garau, C., and Pavan, V. (2018). Evaluating urban quality: Indicators and assessment tools for smart sustainable cities, *Sustainability*, *10*(3), 575.

Giffinger, R., Fertner, C., Kramar, H., Kalasek, R., Pichler-Milanovic, N., and Meijers, E. (2007). *Smart Cities: Ranking of European Medium-Sized Cities*. Vienna: Vienna University of Technology.

Glaeser, E.L., Ponzetto, G.A., and Tobio, K. (2014). Cities, skills and regional change, *Regional Studies*, *48*(1), 7–43.

Glaeser, E.L., Rosenthal, S.S., and Strange, W.C. (2001). Urban economics and entrepreneurship, *Journal of Urban Economics*, *67*(1): 1–14.

Höjer, M., and Wangel, J. (2015). Smart sustainable cities: Definition and challenges, in: *ICT innovations for sustainability*, pp. 333–349. Cham: Springer.

Holden, M. (2006). Urban indicators and the integrative ideals of cities, *Cities*, *23*(3), 170–183.

Holden, M. (2013). Sustainability indicator systems within urban governance: Usability analysis of sustainability indicator systems as boundary objects, *Ecological Indicators*, *32*, 89–96.

Hollands, R.G. (2008). Will the real smart city please stand up? Intelligent, progressive or entrepreneurial? *Cities*, *12*(3), 303–320.

Ibrahim, M., El-Zaart, A., and Adams, C. (2018). Smart sustainable cities roadmap: Readiness for transformation towards urban sustainability, *Sustainable Cities and Society*, *37*, 530–540.

Ivaldi, E., and Ciacci, A. (2020). Measuring level of technological infrastructure in smart cities: A non-compensatory approach, in: Dinçer, H., and Yüksel, S. (eds.), *Strategic Priorities in Competitive Environments in Multidimensional Approaches for Business Success.* Springer in press.

Ivaldi, E., Ciacci, A., and Soliani, R. (2020a). Urban deprivation in Argentina: A POSET analysis, *Papers in Regional Science*. URL: https://doi.org/10.1111/pirs.12555.

Ivaldi E., Penco L., Isola G., and Musso E. (2020b). Smart sustainable cities and the urban knowledge-based economy. A NUTS3 level analysis, *Social indicator Research*. URL: https://doi.org/10.1007/s11205-020-02292-0.

Klopp J., and Petretta D. (2017). The urban sustainable development goal: Indicators, complexity and the politics of measuring cities, *Cities*, *63*, 92–97.

Kramers, A., Wangel, J., and Höjer, M. (2016). Governing the smart sustainable city: The case of Stockholm Royal Seaport, in: *ICT for Sustainability 2016*. Atlantis Press.

Krueger, R., and D. Gibbs. (2008). 'Third Wave' Sustainability? Smart growth and regional development in the USA, *Regional Studies*, *42*(9), 1,263–1,274.

Kummitha, R.K.R. (2019). Smart cities and entrepreneurship: An agenda for future research, *Technological Forecasting and Social Change*, *149*, 119763.

Maggino, F. (2017). Dealing with syntheses in a system of indicators, in: Maggino, F. (ed.), *Complexity in Society: From Indicators Construction to their Synthesis*, pp. 115–137. Cham: Springer.

Martin, C.J., Evans, J., and Karvonen, A. (2018). Smart and sustainable? Five tensions in the visions and practices of the smart-sustainable city in Europe and North America, *Technological Forecasting and Social Change, 133,* 269–278.

Méndez, R., and Moral, S.S. (2011). Spanish cities in the knowledge economy: Theoretical debates and empirical evidence, *European Urban and Regional Studies, 18*(2), 136–155.

Moreno Pires, S., Fidélis T., and Ramos T.B. (2014). Measuring and comparing local sustainable development through common indicators: Constraints and achievements in practice, *ScienceDirect, 39,* 1–9. DOI: 10.1016/j.cities.2014.02.003

Nam, T., and Pardo, T.A. (2011). Conceptualizing smart city with dimensions of technology, people, and institutions, in: *Proceedings of the 12th Annual International Digital Government Research Conference: Digital Government Innovation in Challenging Times*, pp. 282–291. ACM.

OECD environmental outlook to 2050. (2012). *The consequences of inaction OECD Publishing,* URL: http://www.naturvardsverket.se/upload/miljoarbete-i-samhallet/internationellt-miljoarbete/multilateralt/oecd/outolook-2050-oecd.pdf.

Osservatorio Nazionale Smart City ONSC–Anci. (2020). *Vademecum per la città intelligente*, Forum PA Edizioni.

Penco, L., Ivaldi, E., Bruzzi, C., and Musso, E. (2020). Knowledge-based urban environments and entrepreneurship: Inside EU cities, *Cities*, 96. URL: https://doi.org/10.1016/j.cities.2019.102443.

Porter, M.E. (1998). Clusters and the new economics of competition, *Harvard Business Review, 76,* 77–90.

Porter, M.E. (2001). *Clusters of Innovation: Regional Foundations of Competitiveness*. Washington, DC: US Council on Competitiveness.

Santinha, G., Dias, A., Rodrigues, M., Queirós, A., Rodrigues, C., and Rocha, N.P. (2019). How do smart cities impact on sustainable urban growth and on opportunities for entrepreneurship? Evidence from Portugal: The case of Águeda, in: *New Paths of Entrepreneurship Development*, pp. 31–53. Cham: Springer.

Scheller D., and Thörn H. (2018). Governing 'sustainable urban development' through self-build groups and co-housing: The cases of hamburg and gothenburg, *International Journal of Urban and Regional Research, 42*(5), 914–933.

Smigiel, C. (2019). Urban political strategies in times of crisis: A multiscalar perspective on smart cities in Italy, *European Urban and Regional Studies*, *26*(4), 336–348.

Tokody, D., and Schuster, G. (2016). Driving forces behind Smart city implementations—The next smart revolution, *Journal of Emerging research and solutions in ICT*, *1*(2), 1–16.

United Nations. (2015). *United Nations World Urbanization Prospects. The 2014 Revision*. New York: Department of Economic and Social Affairs. URL: http://esa.un.org/unpd/wup/Publications/Files/WUP2014-Report.pdf.

Washburn, D., Sindhu, U., Balaouras, S., Dines, R.A., Hayes, N., and Nelson, L.E. (2009). Helping CIOs understand "smart city" initiatives, *Growth*, *17*(2), 1–17.

Yigitcanlar, T., Baum, S., and Horton, S. (2007). Attracting and retaining knowledge workers in knowledge cities, *Journal of Knowledge Management*, *11*(5), 6–17.

© 2023 World Scientific Publishing Company
https://doi.org/10.1142/9789811257186_bmatter

Index

G

H

I

Ingram Content Group UK Ltd.
Milton Keynes UK
UKHW022030080323
418232UK00002B/50

9 789811 257179